Rigorous Proofs for Riemann Hypothesis, Polignac's and Twin Prime Conjectures in 2020

©Professor Bernhard (Pseudonym) Riemann, viXra, Wednesday 15 January 2020

©*Professor Bernhard (Pseudonym) Riemann, viXra, Wednesday 15 January 2020*

ABSTRACT

Dedicated to my youngest daughter Jelena born 13 weeks early on May 14, 2012. The author openly declares that on January 20, 2020 he was successful in gratefully receiving research grant of AUS $5,000 generously offered by Mrs. Connie Hayes and Mr. Colin Webb to support conduct of ongoing mathematical research. During the relevant research period, he also received AUS $3,250 reimbursement from Q-Pharm for participating in EyeGene Shingles trial commencing on March 10, 2020. Huge thanks to Rodney Williams (Civil Engineer and Mathematician), Tony O'Hagan (Software Engineer and Mathematician), and involved Experts for contributing to the invaluable completed peer reviews required by this paper. **Please generously donate some money to support *John Ting's Self-funded Research Project* located at URL https://jycting.wordpress.com/** RED ALERT: In addition to direct and indirect spin-offs from solving Riemann hypothesis (and explaining two types of Gram points) and Polignac's & Twin prime conjectures [which will require obeying Complex Elementary Fundamental Laws]; we validly expand on tertiary spin-offs which consist of extrapolating Fundamental Laws to include deriving Complex Emergent Fundamental Laws [which will provide a model for explaining the Incompletely Predictable nature of the unprecedented Coronavirus COVID-19 Pandemic declared by World Health Organization on Wednesday March 11, 2020]. This particular subject will constitute the main topic of my next book titled "Coronavirus COVID-19 Pandemic, Gecko, Guinea Pig and Researcher in 2020".

Mathematics for Incompletely Predictable Problems is associated with Incompletely Predictable problems containing Incompletely Predictable entities. Nontrivial zeros and two types of Gram points in Riemann zeta function (or its proxy Dirichlet eta function) together with prime and composite numbers from Sieve of Eratosthenes are Incompletely Predictable entities. Correct and complete mathematical arguments for first key step of converting this function into its continuous format version, and second key step of applying Information-Complexity conservation to this Sieve result in direct spin-offs from first key step consisting of proving Riemann hypothesis (and explaining two types of Gram points), and second key step consisting of proving Polignac's and Twin prime conjectures. ISBN 9781660905768, https://www.amazon.com/-/e/B07T4T

Keywords:
Dirichlet Sigma-Power Laws, Exact and Inexact Dimensional analysis homogeneity, Information-Complexity conservation, Plus Gap 2 Composite Number Continuous Law, Plus-Minus Gap 2 Composite Number Alternating Law, Polignac's & Twin prime conjectures, Riemann hypothesis

1. Introduction

In this research paper, treatise on relevant Mathematics for Incompletely Predictable Problems required to solve Riemann hypothesis and explain the closely related two types of Gram points is outlined first; and to solve Polignac's and Twin prime conjectures is outlined subsequently.

Riemann zeta function	Sieve of Eratosthenes
↓ [Path A option] ↓	↓ [Path A option] ↓
Nontrivial zeros & two types of Gram points	Prime & Composite numbers
↑ [Path B option] ↑	↑ [Path B option] ↑
Dirichlet Sigma-Power Laws	**Dimension (2x - N), N = 2x - ΣPC$_x$-Gap**

Dirichlet Sigma-Power Laws are continuous format version of discrete format Riemann zeta function (or its *proxy* Dirichlet eta function). Sieve of Eratosthenes is a simple ancient algorithm for finding all prime numbers up to any given limit by iteratively marking as composite (i.e., not prime) the multiples of each prime, starting with first prime number 2. Multiples of a given prime are generated as a sequence of numbers starting from that prime, with constant difference between them equal to that prime. Dimension (2x - N) [see Section 8 for further information] *dependently* incorporate prime and composite numbers (and Number '1') whereas Sieve of Eratosthenes directly and indirectly give rise to prime and composite numbers (but not Number '1'). Note that in using the unique Dimension (2x - N) system with N = 2x - ΣPC$_x$-Gap, Dimension (2x - N) when fully expanded is numerically just equal to ΣPC$_x$-Gap since Dimension (2x - N) = 2x - 2x + ΣPC$_x$-Gap = ΣPC$_x$-Gap. *In order to solve Riemann hypothesis (and provide explanations for two types of Gram points), Polignac's and Twin prime conjectures; one could in principle use Path A or Path B option.* Our chosen Path B option requires Mathematics for Incompletely Predictable Problems.

Elements of three complete sets constituted by nontrivial zeros and two types of Gram points together with elements of two complete sets constituted by prime and composite numbers are all classified as Incompletely Predictable entities. Riemann hypothesis (1859) proposed all nontrivial zeros in Riemann zeta function to be located on its critical line. Defined as Incompletely Predictable problem is essential in obtaining the continuous format version of [discrete format] Riemann zeta function dubbed Dirichlet Sigma-Power Law to prove this hypothesis. All of infinite magnitude, nontrivial zeros when geometrically depicted as corresponding Origin intercepts together with two types of Gram points when geometrically depicted as corresponding x- & y-axes intercepts explicitly confirm they intrinsically form relevant component of point-intersections in this function. Defined as Incompletely Predictable problems is essential for these explanations to be correct. Involving proposals that prime gaps and associated sets of prime numbers are infinite in magnitude, Twin prime conjecture (1846) deals with even prime gap 2 thus forming a subset of Polignac's conjecture (1849) which deals with all even prime gaps 2, 4, 6, 8, 10,.... Defined as Incompletely Predictable problems is essential to prove these conjectures using research method dubbed Information-Complexity conservation. Then Mathematics for Incompletely Predictable Problems equates to *sine qua non* defining problems involving Incompletely Predictable entities to be Incompletely Predictable problems achieved by incorporating certain identifiable mathematical steps with this procedure ultimately enabling us to rigorously prove or explain above problems as direct spin-offs.

Refined information on Incompletely Predictable entities of Gram and virtual Gram points: These entities all of infinite magnitude are *dependently* calculated using *complex equation* Riemann zeta function, $\zeta(s)$, or its *proxy* Dirichlet eta function, $\eta(s)$, in critical strip (denoted by $0 < \sigma < 1$) thus forming relevant component of point-intersections in this function. In Figure 2 below, Gram[y=0], Gram[x=0] and Gram[x=0,y=0] points are respectively geometrical x-axis, y-axis and Origin intercepts at critical line (denoted by $\sigma = \frac{1}{2}$). Gram[y=0] and Gram[x=0,y=0] points are respectively synonymous with traditional *'Gram points'* and *nontrivial zeros*. In Figures 3 and 4 below, virtual Gram[y=0] and virtual Gram[x=0] points are respectively geometrical x-axis and y-axis intercepts at non-critical lines (denoted by $\sigma \neq \frac{1}{2}$). Virtual Gram[x=0,y=0] points do not exist.

Refined information on Incompletely Predictable entities of prime and composite numbers: These entities all of infinite magnitude are *dependently* computed (respectively) directly and indirectly using *complex algorithm* Sieve of Eratosthenes. The statement Set **N** is formed by Set **P**, Set **C** and Number '1' can easily be validated to be true. Denote $\mathbb{C}$ to be uncountable complex numbers, **R** to be uncountable real numbers, **Q** to be countable rational numbers or roots [of non-zero polynomials], **R**–**Q** to be uncountable irrational numbers, **A** to be countable algebraic numbers, **R**–**A** to be uncountable transcendental numbers, **Z** to be countable integers, **W** to be countable whole numbers, **N** to be countable natural numbers, **E** to be countable even numbers, **O** to be countable odd numbers, **P** to be countable prime numbers, and **C** to be countable composite numbers. **A** are $\mathbb{C}$ (including **R**) that are countable rational or irrational roots. We have (i) Set **N** = Set **E** + Set **O**, (ii) Set **N** = Set **P** + Set **C** + Number '1', (iii) Set **A** = Set **Q** + Set **irrational roots**, and (iv) Set **N** $\subset$ Set **W** $\subset$ Set **Z** $\subset$ Set **Q** $\subset$ Set **R** $\subset$ Set $\mathbb{C}$. Then Set **R**–**Q** = Set **irrational roots** + Set **R**–**A**.

Cardinality of a given set: With increasing size, arbitrary Set **X** can be countable finite set (CFS), countable infinite set (CIS) or uncountable infinite set (UIS). Cardinality of Set **X**, |**X**|, measures "number of elements" in Set **X**. E.g. Set **negative Gram[y=0] point** has CFS of negative Gram[y=0] point with |**negative Gram[y=0] point**| = 1, Set **even P** has CFS of even **P** with |**even P**| = 1, Set **N** has CIS of **N** with |**N**| = $\aleph_0$, and Set **R** has UIS of **R** with |**R**| = $\mathfrak{c}$ (cardinality of the continuum).

Differentiation of terms "Incompletely Predictable" versus "Completely Predictable": Set **N** = Set **E** + Set **O**. The two subsets of even and odd numbers are "Independent" and "Completely Predictable". Examples: The even number after 2,984 viz. 2,984 / 2 = 1,492nd even number is [easily] calculated independently using *simple algorithm* to be 2,984+2 = 2,986 viz. 2,986 / 2 = 1,493rd even number. The odd number after 2,985 viz. (2,985+1) / 2 = 1,493rd odd number is [easily] calculated independently using *simple algorithm* to be 2,985+2 = 2,987 viz. (2,987+1) / 2 = 1,494th odd number. Set **N** = Set **P** + Set **C** + Number '1'. The two subsets of prime and composite numbers are "Dependent" and "Incompletely Predictable". Example: The sixth prime number '13' [after fifth prime number '11'] is [not easily] computed dependently using *complex algorithm* from scratch via: 2 is 1st prime number, 3 is 2nd prime number, 4 is 1st composite number, 5 is 3rd prime number, 6 is 2nd composite number, 7 is 4th prime number, 8 is 3rd composite number, 9 is 4th composite number, 10 is 5th composite number, 11 is 5th prime number, 12 is 6th composite number, and our desired 13 is 6th prime number.

Formal definitions for Completely Predictable (CP) entities and Incompletely Predictable (IP) entities: In this paper, the word "number" [singular noun] or "numbers" [plural noun] in reference to prime & composite numbers, nontrivial zeros & two types of Gram points can interchangeably be replaced with the word "entity" [singular noun] or "entities" [plural noun]. Respectively, an IP (CP) number is locationally defined as a number whose position is *dependently* (*independently*) determined by complex (simple) calculations using complex (simple) equation or algorithm with (without) needing to know related positions of all preceding numbers in neighborhood. Simple properties are inferred from a sentence such as "This simple equation or algorithm by itself will intrinsically incorporate actual location [and actual positions] of all CP numbers". Solving CP problems with simple properties amendable to *simple* treatments using *usual* mathematical tools such as Calculus result in 'Simple Elementary Fundamental Laws'-based solutions. Complex properties, or "meta-properties", are inferred from a sentence such as "This complex equation or algorithm by itself will intrinsically incorporate actual location [but not actual positions] of all IP numbers". Solving IP problems with complex properties amendable to *complex* treatments using *unusual* mathematical tools such as Information-Complexity conservation, exact and inexact Dimensional analysis homogeneity as well as using *usual* mathematical tools such as Calculus result in 'Complex Elementary Fundamental Laws'-based solutions.

Based on Mathematics for Incompletely Predictable Problems, we compare and contrast CP entities (obeying Simple Elementary Fundamental Laws) against IP entities (obeying Complex Elementary Fundamental Laws) using the following examples:
(I) **E** are CP entities constituted by CIS of **Q** 2, 4, 6, 8, 10, 12....
(II) **O** are CP entities constituted by CIS of **Q** 1, 3, 5, 7, 9, 11....
(III) **P** are IP entities constituted by CIS of **Q** 2, 3, 5, 7, 11, 13....
(IV) **C** are IP entities constituted by CIS of **Q** 4, 6, 8, 9, 10, 12....
(V) With values traditionally given by parameter t, nontrivial zeros in Riemann zeta function are IP entities constituted by CIS of **R**–**A** [rounded off to six decimal places]: 14.134725, 21.022040, 25.010858, 30.424876, 32.935062, 37.586178,....
(VI) Traditional 'Gram points' (or Gram[y=0] points) are x-axis intercepts with choice of index 'n' for 'Gram points' historically chosen such that first 'Gram point' [by convention at n = 0] corresponds to the t value which is larger than (first) nontrivial zero located at t = 14.134725. 'Gram points' are IP entities constituted by CIS of **R**–**A** [rounded off to six decimal places] with the first six given at n = -3, t = 0; at n = -2, t = 3.436218; at n = -1, t = 9.666908; at n = 0, t = 17.845599; at n = 1, t = 23.170282; at n = 2, t = 27.670182.

Denoted by parameter t; nontrivial zeros, 'Gram points' and Gram[x=0] points all belong to well-defined CIS of **R**–**A** which will twice obey the relevant location definition [in CIS of **R**–**A** themselves and in CIS of numerical digits after decimal point of each **R**–**A**]. First and only negative 'Gram point' (at n = -3) is obtained by substituting CP t = 0 resulting in $\zeta(\frac{1}{2} + it)$ = $\zeta(\frac{1}{2})$ = -1.4603545, a **R**–**A** number [rounded off to seven decimal places] calculated as a limit similar to limit for Euler-Mascheroni constant or Euler gamma with its precise (1st) position only determined by computing positions of all preceding (nil) 'Gram point' in this case. '0' and '1' are special numbers being neither **P** nor **C** as they represent nothingness (zero) and wholeness (one). In this setting, the ideas of (i) having factors for '0' and '1', or (ii) treating '0' and '1' as CP or IP numbers, is meaningless. All entities derived from well-defined simple/complex algorithms or equations are "dual numbers" as they can be simultaneously depicted as CP and IP numbers. For instance, **Q** '2' as **P** (& **E**), '97' as **P** (& **O**), '98' as **C** (& **E**), '99' as **C** (& **O**); CP '0' values in x=0, y=0 & simultaneous x=0,y=0 associated with various IP t values in $\zeta(s)$.

1.1 Algebraic number theory versus Analytic number theory

Set **P** $\subset$ Set **Z** $\subset$ Set **Q**. Gaussian rationals, and Gaussian integers are complex numbers whose real and imaginary parts are (respectively) both rational numbers, and integer numbers. Gaussian primes are Gaussian integers z = a + bi satisfying one of the following properties.

1. If both a and b are nonzero, then a+bi is a Gaussian prime iff $a^2 + b^2$ is an ordinary prime [whereby iff is the written abbreviation for 'if and only if'].

2. If a = 0, then bi is a Gaussian prime iff |b| is an ordinary prime and |b| = 3 (mod 4).

3. If b = 0, then a is a Gaussian prime iff |a| is an ordinary prime and |a| = 3 (mod 4).

Prime numbers which are also Gaussian primes are 3, 7, 11, 19, 23, 31, 43,.... In Eq. (1) below, we noted that the equivalent Euler product formula with product over prime numbers [instead of summation over natural numbers] faithfully represent Riemann zeta function, $\zeta(s)$. Eq. (2) below is Riemann's functional equation involving transcendental number π (= 3.14159...). With denominators on the left involving odd numbers and named after Gottfried Leibniz, Leibniz formula for π states that $\frac{1}{1} - \frac{1}{3} + \frac{1}{5} - \frac{1}{7} + \frac{1}{9} - \cdots = \frac{\pi}{4}$. Expression $\zeta(2) = \frac{1}{1^2} + \frac{1}{2^2} + \frac{1}{3^2} + \cdots = \frac{\pi^2}{6}$ $\approx 1.644934066848226436 47$ involves $\pi \implies$ division concerning two unrelated transcendental (irrational) numbers as $\frac{\zeta(2)}{\pi^2}$ will interestingly result in rational number $\frac{1}{6}$.

Algebraic number theory is loosely defined to deal with new number systems involving Completely Predictable or Incompletely Predictable entities such as even & odd numbers, prime & composite numbers, p-adic numbers, Gaussian primes, Gaussian rationals & integers, and complex numbers. A p-adic number is an extension of the field of rationals such that congruences modulo powers of a fixed prime number p are related to proximity in so-called "p-adic metric". The extension is achieved by an alternative interpretation of concept of "closeness" or absolute value viz. p-adic numbers are considered to be close when their difference is divisible by a high power of p: the higher the power, the closer they are. This property enables p-adic numbers to encode congruence information in a way that turns out to have powerful applications in number theory including, for example, attacking certain Diophantine equations and in the famous proof of Fermat's Last Theorem by English mathematician Sir Andrew John Wiles in 1995.

Analytic number theory is loosely defined to deal with functions of a complex variable such as Riemann zeta function [containing nontrivial zeros & two types of Gram points] and other L-functions. Study of prime numbers, complex numbers & π being braided together in a pleasing trio is usefully visualized to be located at intersection of this two main branches of number theory. We loosely separate our relatively elementary proof for Riemann hypothesis & relatively elementary explanations for two types of Gram points to belong to Analytic number theory, and our relatively elementary proofs for Polignac's & Twin prime conjectures [expectedly associated with paucity of functions of a complex variable] to belong to Algebraic number theory.

Indirect spin-offs from solving Riemann hypothesis are often stated as "With this one solution, we have proven five hundred theorems or more at once". This apply to many important theorems in Number theory (mostly on prime numbers) that rely on properties of Riemann zeta function such as where trivial and nontrivial zeros are / are not located. A classical example is resulting absolute and full delineation of prime number theorem, which relates to prime counting function. This function, usually denoted by $\pi(x)$, is defined as the number of prime numbers $\leqslant$ x. Public-key cryptography that is widely required for financial security in E-Commerce traditionally depend on solving the difficult problem of factoring prime numbers for astronomically large numbers. The intrinsic "Incompletely Predictable" property present in prime numbers, composite numbers, nontrivial zeros and two types of Gram points can never be altered to "Completely Predictable" property. For this stated reason, it is a mathematical impossibility that providing rigorous proofs such as for Riemann hypothesis will in principle ever result in crypto-apocalypse. However, utilizing parallel computing (more than seriel computing), fast supercomputers and the far-more-powerful quantum computers would theoretically allow solving difficult factorization problem in quick time. This will result in less secure encryption and decryption. Then using, for instance, quantum cryptography that relies on principles of quantum mechanics to encrypt data and transmit it in a way that cannot be hacked will combat this issue.

Proposed by German mathematician Bernhard Riemann (September 17, 1826 – July 20, 1866) in 1859, Riemann hypothesis is mathematical statement on $\zeta(s)$ that critical line denoted by σ = $\frac{1}{2}$ contains complete Set **nontrivial zeros** with |**nontrivial zeros**| = $\aleph_0$. Alternatively, this hypothesis is geometrical statement on $\zeta(s)$ that generated curves when $\sigma = \frac{1}{2}$ contain complete Set **Origin intercepts** with $|\text{Origin intercepts}| = \aleph_0$.

$$\zeta(s) = \frac{e^{(\ln(2\pi)-1-\frac{\gamma}{2})s}}{2(s-1)\Gamma(1+\frac{s}{2})} \Pi_\rho \left(1 - \frac{s}{\rho}\right) e^{\frac{s}{\rho}} = \pi^{\frac{s}{2}} \frac{\Pi_\rho \left(1 - \frac{s}{\rho}\right)}{2(s-1)\Gamma\left(1+\frac{s}{2}\right)}$$

Depicted in full and abbreviated version, Hadamard product above is infinite product expansion of $\zeta(s)$ based on Weierstrass's factorization theorem displaying a simple pole at s = 1. It contains both trivial & nontrivial zeros indicating their common origin from $\zeta(s)$. Set **trivial zeros** occurs at $\sigma = $ -2, -4, -6, -8, -10,..., ∞ with $|\text{trivial zeros}| = \aleph_0$ due to Γ function term in denominator. Nontrivial zeros occur at $s = \rho$ with γ denoting Euler-Mascheroni constant.

Remark 1.1. Confirming first 10,000,000,000,000 nontrivial zeros location on critical line implies but does not prove Riemann hypothesis to be true.

Locations of first 10,000,000,000,000 nontrivial zeros on critical line have previously been computed to be correct. Hardy in 1914[1], and with Littlewood in 1921[2], showed infinite nontrivial zeros on critical line by considering moments of certain functions related to $\zeta(s)$. This discovery cannot constitute rigorous proof for Riemann hypothesis because they have not exclude theoretical existence of nontrivial zeros located away from this line.

1.2 Exact & inexact Dimensional analysis homogeneity for equations & inequations

Respectively for 'base quantities' such as *length*, *mass* and *time*; their fundamental SI 'units of measurement' meter (m) is defined as distance travelled by light in vacuum for time interval 1/299 792 458 s with speed of light c = 299,792,458 ms^{-1}, kilogram (kg) is defined by taking fixed numerical value Planck constant h to be 6.626 070 15 X 10^{-34} Joules·second (Js) [whereby Js is equal to kgm^{2}s^{-1}] and second (s) is defined in terms of ΔvCs = $\Delta(^{133}\text{Cs})_{hfs}$ = 9,192,631,770 s^{-1}. Derived SI units such as J and ms^{-1} respectively represent 'base quantities' *energy* and *velocity*. The word 'dimension' is commonly used to indicate all those mentioned 'units of measurement' in well-defined equations.

Dimensional analysis (DA) is an analytic tool with DA homogeneity and non-homogeneity (respectively) denoting valid and invalid equation occurring when 'units of measurements' for 'base quantities' are "balanced" and "unbalanced" across both sides of the equation. E.g. equation 2 m + 3 m = 5 m is valid and equation 2 m + 3 kg = 5 mkg is invalid (respectively) manifesting DA homogeneity and non-homogeneity.

Remark 1.2. We can validly apply exact and inexact Dimensional analysis homogeneity to well-defined equations and inequations.

Let (2n) and (2n-1) be 'base quantities' in our derived Dirichlet Sigma-Power Laws formatted in simplest forms as equations and inequations. E.g. DA on exponent $\frac{1}{2}$ in $(2n)^{\frac{1}{2}}$ when depicted in simplest form is correct but DA on exponent $\frac{1}{4}$ in equivalent $(2^2 n^2)^{\frac{1}{4}}$ *not* depicted in simplest form is incorrect. Fractional exponents as 'units of measurement' given by $(1-\sigma)$ for equations and $(\sigma+1)$ for inequations when $\sigma = \frac{1}{2}$ coincide with exact DA homogeneity[1]; and $(1-\sigma)$ for equations and $(\sigma+1)$ for inequations when $\sigma \neq \frac{1}{2}$ coincide with inexact DA homogeneity[2]. Respectively for equations and inequations, exact DA homogeneity at $\sigma = \frac{1}{2}$ denotes $\sum$(all fractional exponents) as $2(1-\sigma)$ and $2(\sigma+1)$ equates to ["exact"] whole number '1' and '3'; and inexact DA homogeneity at $\sigma \neq \frac{1}{2}$ denotes $\sum$(all fractional exponents) as $2(1-\sigma)$ and $2(\sigma+1)$ equates to ["inexact"] fractional number '$\neq$1' and '$\neq$3'.

Footnote 1, 2: Exact and inexact DA homogeneity occur in Dirichlet Sigma-Power Laws as equations or inequations for Gram[y=0] points, Gram[x=0] points and nontrivial zeros. *Law of Continuity* is a heuristic principle *whatever succeed for the finite, also succeed for the infinite*. Then these Laws which inherently manifest themselves on finite and infinite time scale should "succeed for the finite, also succeed for the infinite".

Outline of proof for Riemann hypothesis. To simultaneously satisfy two mutually inclusive conditions: I. *With rigid manifestation of exact DA homogeneity*, Set **nontrivial zeros** with |**nontrivial zeros**| = $\aleph_0$ is located on critical line (viz. $\sigma = \frac{1}{2}$) when $2(1-\sigma)$ [or $2(\sigma+1)$] as $\sum$(all fractional exponents) = whole number '1' [or '3'] in Dirichlet Sigma-Power Law[3] as equation [or inequation]. II. *With rigid manifestation of inexact DA homogeneity*, Set **nontrivial zeros** with |**nontrivial zeros**| = $\aleph_0$ is not located on non-critical lines (viz. $\sigma \neq \frac{1}{2}$) when $2(1-\sigma)$ [or $2(\sigma+1)$] as $\sum$(all fractional exponents) = fractional number '$\neq 1$' [or '$\neq 3$'] in Dirichlet Sigma-Power Law[3] as equation [or inequation].

Footnote 3: Derived from original $\eta(s)$ (*proxy* for $\zeta(s)$) as equation or inequation, this Law symbolizes end-result proof on Riemann hypothesis.

Riemann hypothesis mathematical foot-prints. Six identifiable steps to prove Riemann hypothesis: *Step 1* Use $\eta(s)$, proxy for $\zeta(s)$, in critical strip. *Step 2* Apply Euler formula to $\eta(s)$. *Step 3* Obtain "simplified" Dirichlet eta function which intrinsically incorporates *actual location [but not actual positions]* of all nontrivial zeros[4]. *Step 4* Apply Riemann integral to "simplified" Dirichlet eta function in discrete (summation) format. *Step 5* Obtain Dirichlet Sigma-Power Law in continuous (integral) format as equation or inequation. *Step 6* Note exact and inexact DA homogeneity associated with their fractional exponents.

Footnote 4: Respectively Gram[y=0] points, Gram[x=0] points and nontrivial zeros are Incompletely Predictable entities with actual positions determined by setting $\sum Im\{\eta(s)\} = 0$, $\sum Re\{\eta(s)\} = 0$ and $\sum ReIm\{\eta(s)\} = 0$ to *dependently* calculate relevant positions of all preceding entities in neighborhood. Respectively actual location of Gram[y=0] points, Gram[x=0] points and nontrivial zeros; and virtual Gram[y=0] points, virtual Gram[x=0] points and "absent" nontrivial zeros occur precisely at $\sigma = \frac{1}{2}$; and $\sigma \neq \frac{1}{2}$.

2. Riemann zeta and Dirichlet eta functions

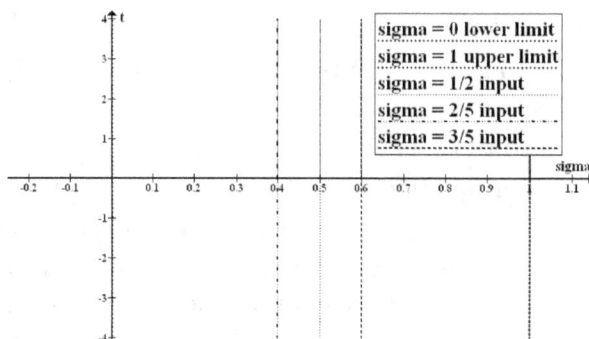

FIGURE 1. INPUT for $\sigma = \frac{1}{2}$, $\frac{2}{5}$, and $\frac{3}{5}$. $\zeta(s)$ has countable infinite set of Completely Predictable trivial zeros at $\sigma =$ all negative even numbers and countable infinite set of Incompletely Predictable nontrivial zeros at $\sigma = \frac{1}{2}$ for various t values.

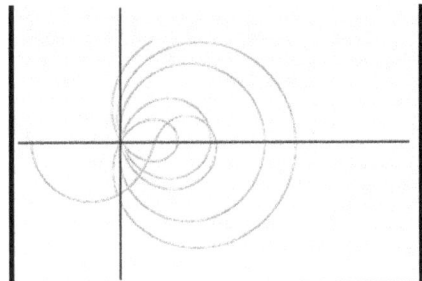

FIGURE 2. OUTPUT for $\sigma = \frac{1}{2}$. Schematically depicted polar graph of $\zeta(\frac{1}{2} + \imath t)$ plotted along critical line for real values of t running from 0 to 34, horizontal axis: $Re\{\zeta(\frac{1}{2} + \imath t)\}$, and vertical axis: $Im\{\zeta(\frac{1}{2} + \imath t)\}$. There are presence of Origin intercepts which are totally absent in Figures 3 and 4 [with identical axes definitions].

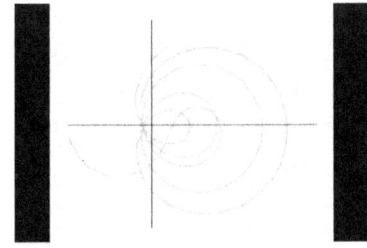

FIGURE 3. OUTPUT for $\sigma = \frac{2}{5}$.

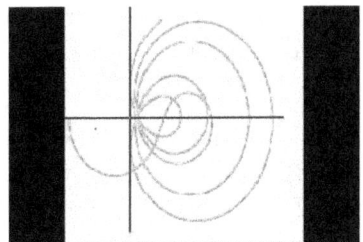

FIGURE 4. OUTPUT for $\sigma = \frac{3}{5}$.

An L-function consists of a Dirichlet series with a functional equation and an Euler product. Examples of L-functions come from modular forms, elliptic curves, number fields, and Dirichlet characters, as well as more generally from automorphic forms, algebraic varieties, and Artin representations. They form an integrated component of 'L-functions and Modular Forms Database' (LMFDB) with far-reaching implications. In perspective, $\zeta(s)$ is the simplest example of an L-function. It is a function of complex variable s ($= \sigma \pm \imath t$) that analytically continues sum of infinite series $\zeta(s) = \sum_{n=1}^{\infty} \frac{1}{n^s} = \frac{1}{1^s} + \frac{1}{2^s} + \frac{1}{3^s} + \cdots$. The common convention is to write s as $\sigma + \imath t$ with $\imath = \sqrt{-1}$, and σ & t real. Valid for $\sigma > 0$, we write $\zeta(s)$ as $Re\{\zeta(s)\} + \imath Im\{\zeta(s)\}$ and note that $\zeta(\sigma + \imath t)$ when $0 < t < +\infty$ is the complex conjugate of $\zeta(\sigma - \imath t)$ when $-\infty < t < 0$.

Also known as alternating zeta function, $\eta(s)$ must act as *proxy* for $\zeta(s)$ in critical strip (viz. $0 < \sigma < 1$) containing critical line (viz. $\sigma = \frac{1}{2}$) because $\zeta(s)$ only converges when $\sigma > 1$. This implies $\zeta(s)$ is undefined to left of this region in critical strip which then requires $\eta(s)$ representation instead. They are related to each other as $\zeta(s) = \gamma \cdot \eta(s)$ with proportionality factor $\gamma = \frac{1}{(1 - 2^{1-s})}$ and $\eta(s) = \sum_{n=1}^{\infty} \frac{(-1)^{n+1}}{n^s} = \frac{1}{1^s} - \frac{1}{2^s} + \frac{1}{3^s} - \cdots$.

$$\zeta(s) = \sum_{n=1}^{\infty} \frac{1}{n^s} \tag{1}$$

$$= \frac{1}{1^s} + \frac{1}{2^s} + \frac{1}{3^s} + \cdots$$

$$= \Pi_{p\ prime} \frac{1}{(1-p^{-s})}$$

$$= \frac{1}{(1-2^{-s})} \cdot \frac{1}{(1-3^{-s})} \cdot \frac{1}{(1-5^{-s})} \cdot \frac{1}{(1-7^{-s})} \cdot \frac{1}{(1-11^{-s})} \cdots \frac{1}{(1-p^{-s})} \cdots$$

Eq. (1) is defined for only $1 < \sigma < \infty$ region where $\zeta(s)$ is absolutely convergent. There are no zeros located here. In Eq. (1), equivalent Euler product formula with product over prime numbers [instead of summation over natural numbers] can also represent $\zeta(s)$.

$$\zeta(s) = 2^s \pi^{s-1} \sin\left(\frac{\pi s}{2}\right) \cdot \Gamma(1-s) \cdot \zeta(1-s) \tag{2}$$

With $\sigma = \frac{1}{2}$ as symmetry line of reflection, Eq. (2) is Riemann's functional equation valid for $-\infty < \sigma < \infty$. It can be used to find all trivial zeros on horizontal line at $\imath t = 0$ occurring when $\sigma = $ -2, -4, -6, -8, -10,..., ∞ whereby $\zeta(s) = 0$ because factor $\sin(\frac{\pi s}{2})$ vanishes. Γ is gamma function, an extension of factorial function [a product function denoted by ! notation whereby $n! = n(n-1)(n-2)\ldots(n-(n-1))$] with its argument shifted down by 1, to real and complex numbers. That is, if n is a positive integer, $\Gamma(n) = (n-1)!$

$$\zeta(s) = \frac{1}{(1-2^{1-s})} \sum_{n=1}^{\infty} \frac{(-1)^{n+1}}{n^s} \tag{3}$$

$$= \frac{1}{(1-2^{1-s})} \left(\frac{1}{1^s} - \frac{1}{2^s} + \frac{1}{3^s} - \cdots \right)$$

Eq. (3) is defined for all $\sigma > 0$ values except for simple pole at $\sigma = 1$. As alluded to above, $\zeta(s)$ without $\frac{1}{(1-2^{1-s})}$ viz. $\sum_{n=1}^{\infty} \frac{(-1)^{n+1}}{n^s}$ is $\eta(s)$. It is a holomorphic function of s defined by analytic continuation and is mathematically defined at $\sigma = 1$ whereby analogous trivial zeros with presence only for $\eta(s)$ [but not for $\zeta(s)$] on vertical straight line $\sigma = 1$ are found at $s = 1 \pm \imath \frac{2\pi k}{\ln(2)}$ where k = 1, 2, 3, 4, 5, ..., ∞.

Figure 1 above depict complex variable s ($= \sigma \pm \imath t$) as INPUT with x-axis denoting real part Re{s} associated with σ, and y-axis denoting imaginary part Im{s} associated with t. Figures 2, 3 and 4 above respectively depict $\zeta(s)$ as OUTPUT for real values of t running from 0 to 34 at $\sigma = \frac{1}{2}$ (critical line), $\sigma = \frac{2}{5}$ (non-critical line), and $\sigma = \frac{3}{5}$ (non-critical line) with x-axis denoting real part Re{$\zeta(s)$} and y-axis denoting imaginary part Im{$\zeta(s)$}. There are infinite types-of-spirals possibilities associated with each σ value arising from all infinite σ values in critical strip. Mathematically proving all nontrivial zeros location on critical line as denoted by solitary $\sigma = \frac{1}{2}$ value equates to geometrically proving all Origin intercepts occurrence at solitary $\sigma = \frac{1}{2}$ value. Both result in rigorous proof for Riemann hypothesis.

3. Prerequisite lemma, corollary and propositions for Riemann hypothesis

Original equation $\eta(s)$, *proxy* for $\zeta(s)$, is treated as unique mathematical object with key properties and behaviors. Containing all x-axis, y-axis and Origin intercepts, it will intrinsically incorporate *actual location [but not actual positions]* of all Gram[y=0] points, Gram[x=0] points and nontrivial zeros. Proofs on lemma, corollary and propositions on nontrivial zeros depict exact and inexact DA homogeneity in both derived equation and inequation. Parallel procedure on Gram[y=0] and Gram[x=0] points in Section 5 below depict exact and inexact DA homogeneity in similarly derived equations and inequations.

Lemma 3.1. "Simplified" Dirichlet eta function is derived directly from Dirichlet eta function with Euler formula application and it will intrinsically incorporate actual location [but not actual positions] of all nontrivial zeros.

Proof. Denote complex number ($\mathbb{C}$) as z = x + $\imath$·y. Then z = Re(z) + $\imath$·Im(z) with Re(z) = x and Im(z) = y; modulus of z, $|z| = \sqrt{Re(z)^2 + Im(z)^2} = \sqrt{x^2 + y^2}$; and $|z|^2 = x^2 + y^2$.

Euler formula is commonly stated as $e^{\imath x} = \cos x + \imath \cdot \sin x$. Euler identity (where $x = \pi$) is $e^{\imath \pi} = \cos \pi + \imath \cdot \sin \pi = -1 + 0$ [or stated as $e^{\imath \pi} + 1 = 0$]. The n^s of $\zeta(s)$ is expanded to $n^s = n^{(\sigma + \imath t)} = n^{\sigma} e^{t \ln(n) \cdot \imath}$ since $n^t = e^{t \ln(n)}$. Apply Euler formula to n^s result in $n^s = n^{\sigma}(\cos(t \ln(n)) + \imath \cdot \sin(t \ln(n)))$. This is written in trigonometric form [designated by short-hand notation $n^s(Euler)$] whereby n^{σ} is modulus and $t \ln(n)$ is polar angle (argument).

Apply $n^s(Euler)$ to Eq. (1). Then $\zeta(s) = Re\{\zeta(s)\} + \imath \cdot Im\{\zeta(s)\}$ with
$Re\{\zeta(s)\} = \sum_{n=1}^{\infty} n^{-\sigma} \cos(t \ln(n))$ and $Im\{\zeta(s)\} = \sum_{n=1}^{\infty} n^{-\sigma} \sin(t \ln(n))$. As Eq. (1) is defined only for $\sigma > 1$ where zeros never occur, we will not carry out further treatment here.

Apply $n^s(Euler)$ to Eq. (3). Then $\zeta(s) = \gamma \cdot \eta(s) = \gamma \cdot [Re\{\eta(s)\} + \imath \cdot Im\{\eta(s)\}]$ with
$$Re\{\eta(s)\} = \sum_{n=1}^{\infty}((2n-1)^{-\sigma} \cos(t \ln(2n-1)) - (2n)^{-\sigma} \cos(t \ln(2n)));$$
$$Im\{\eta(s)\} = \sum_{n=1}^{\infty}((2n)^{-\sigma} \sin(t \ln(2n)) - (2n-1)^{-\sigma} \sin(t \ln(2n-1)));$$
and proportionality factor $\gamma = \dfrac{1}{(1 - 2^{1-s})}$.

Complex number s in critical strip is designated by s = $\sigma + \imath t$ for $0 < t < +\infty$ and s = $\sigma - \imath t$ for $-\infty < t < 0$. Nontrivial zeros equating to $\zeta(s) = 0$ give rise to our desired $\eta(s) = 0$. Modulus of $\eta(s)$, $|\eta(s)|$, is defined as $\sqrt{(Re\{\eta(s)\})^2 + (Im\{\eta(s)\})^2}$ with $|\eta(s)|^2 = (Re\{\eta(s)\})^2 + (Im\{\eta(s)\})^2$. Mathematically $|\eta(s)| = |\eta(s)|^2 = 0$ is an unique condition giving rise to $\eta(s) = 0$ occurring only when $Re\{\eta(s)\} = Im\{\eta(s)\} = 0$ as any non-zero values for $Re\{\eta(s)\}$ and/or $Im\{\eta(s)\}$ will always result in $|\eta(s)|$ and $|\eta(s)|^2$ having non-zero values. Important implication is that sum of $Re\{\eta(s)\}$ and $Im\{\eta(s)\}$ equating to zero [given by Eq. (4)] must always hold when $|\eta(s)| = |\eta(s)|^2 = 0$ and consequently $\eta(s) = 0$.

$$\sum ReIm\{\eta(s)\} = Re\{\eta(s)\} + Im\{\eta(s)\} = 0 \qquad (4)$$

In principle, advocating for existence of theoretical s values leading to non-zero values in $Re\{\eta(s)\}$ and $Im\{\eta(s)\}$ depicted as possibility $+Re\{\eta(s)\} = -Im\{\eta(s)\}$ or $-Re\{\eta(s)\} = +Im\{\eta(s)\}$ could satisfy Eq. (4). This reverse implication is not necessarily true as these s values will not result in $|\eta(s)| = |\eta(s)|^2 = 0$. In any event, we need not consider these two possibilities since solving Riemann hypothesis involves nontrivial zeros defined by $\eta(s) = 0$ with non-zero values in $Re\{\eta(s)\}$

and/or $Im\{\eta(s)\}$ not compatible with $\eta(s) = 0$.

Riemann hypothesis proposed all nontrivial zeros to be located on critical line. This location is conjectured to be uniquely associated with presence of exact DA homogeneity in derived equation and inequation of Dirichlet Sigma-Power Law with Eq. (4) intrinsically incorporated into this Law as the $\eta(s) = 0$ definition for nontrivial zeros equates to Eq. (4).

Apply trigonometry identity $\cos(x) - \sin(x) = \sqrt{2}\sin\left(x + \frac{3}{4}\pi\right)$ to $Re\{\eta(s)\} + Im\{\eta(s)\}$ to get Eq. (5) with terms in last line built by mixture of terms from $Re\{\eta(s)\}$ and $Im\{\eta(s)\}$.

$$\sum ReIm\{\eta(s)\} = \sum_{n=1}^{\infty}[(2n-1)^{-\sigma}\cos(t\ln(2n-1)) - (2n-1)^{-\sigma}\sin(t\ln(2n-1))$$
$$- (2n)^{-\sigma}\cos(t\ln(2n)) + (2n)^{-\sigma}\sin(t\ln(2n))]$$
$$= \sum_{n=1}^{\infty}[(2n-1)^{-\sigma}\sqrt{2}\sin(t\ln(2n-1) + \frac{3}{4}\pi) - (2n)^{-\sigma}\sqrt{2}\sin(t\ln(2n) + \frac{3}{4}\pi)] \quad (5)$$

When depicted in terms of Eq. (4), Eq. (5) becomes

$$\sum_{n=1}^{\infty}(2n)^{-\sigma}\sqrt{2}\sin(t\ln(2n) + \frac{3}{4}\pi) = \sum_{n=1}^{\infty}(2n-1)^{-\sigma}\sqrt{2}\sin(t\ln(2n-1) + \frac{3}{4}\pi)$$

$$\sum_{n=1}^{\infty}(2n)^{-\sigma}\sqrt{2}\sin(t\ln(2n) + \frac{3}{4}\pi) - \sum_{n=1}^{\infty}(2n-1)^{-\sigma}\sqrt{2}\sin(t\ln(2n-1) + \frac{3}{4}\pi) = 0 \quad (6)$$

Eq. (6) in discrete (summation) format is a non-Hybrid integer sequence equation – see Appendix C. $\eta(s)$ calculations for all σ values result in infinitely many non-Hybrid integer sequence equations for $0<\sigma<1$ critical strip region of interest with n = 1, 2, 3, 4, 5,..., ∞ as discrete integer number values, or n = 1 to ∞ as continuous real numbers values with Riemann integral application. These equations will geometrically represent entire plane of critical strip, thus (at least) allowing our proposed proof to be of a complete nature.

Eq. (6) being the "simplified" Dirichlet eta function derived directly from $\eta(s)$ will intrinsically incorporate *actual location [but not actual positions]* of all nontrivial zeros. *The proof is now complete for Lemma 3.1*□.

Proposition 3.2. Dirichlet Sigma-Power Law in continuous (integral) format given as equation and inequation can both be derived directly from "simplified" Dirichlet eta function in discrete (summation) format with Riemann integral application. [Note: Dirichlet Sigma-Power Law in continuous (integral) format refers to the end-product obtained from "first key step of converting Riemann zeta function into its continuous format version".]

Proof. In Calculus, integration is reverse process of differentiation viewed geometrically as numerical "total area value" solution enclosed by curve of function and x-axis. Apply definite integral I between limits (or points) a and b is to compute its value when $\Delta x \longrightarrow 0$, i.e. $I = \lim_{\Delta x \to 0}\sum_{i=1}^{n}f(x_i)\Delta x_i = \int_{a}^{b}f(x)dx$. This is Riemann integral of function f(x) in interval [a, b] where a<b. Apply Riemann integral to "simplified" Dirichlet eta function in [$\Delta x \longrightarrow 1$] discrete (summation) format which intrinsically incorporates *actual location [but not actual positions]* of all nontrivial zeros criterion to obtain Dirichlet Sigma-Power Law in [$\Delta x \longrightarrow 0$] continuous (integral) format with the later validly representing the former. Then Dirichlet Sigma-Power

Law will also fullfil this criterion. Due to resemblance to power law functions in σ from s = $\sigma + it$ being exponent of a power function n^σ, logarithm scale use, and harmonic $\zeta(s)$ series connection in Zipf's law; we elect to call this Law by its given name. A characteristic and crucial part of this Law is its exact formula expression in usual mathematical language [$y = f(x_1, x_2)$ format description for a 2-variable function with $(2n)$ and $(2n-1)$ as 'base quantities'] consist of $y = f(t, \sigma)$ with discrete n = 1, 2, 3, 4, 5,..., ∞ or continuous n = 1 to ∞; $-\infty < t < +\infty$; and $0 < \sigma < 1$.

A proper integral is a definite integral which has neither limit a or b infinite and from which the integrand does not approach infinity at any point in the range of integration. Only a proper integral will have its [solitary] combined +ve (above x-axis) and -ve (below x-axis) non-zero numerical "total area value" solution successfully computed from applying Riemann integral. An improper integral is a definite integral that has either or both limits a and b infinite or an integrand that approaches infinity at one or more points in the range of integration.

The resulting Dirichlet Sigma-Power Law, being improper integral (with lower limit a = 1 and upper limit b = ∞) obtained from [validly] applying Riemann integral to "simplified" Dirichlet eta function, will [expectedly] have its [multiple] +ve (above x-axis) minus -ve (below x-axis) numerical "net area value" solutions successfully computed – see Propositions 3.3 and 3.4 below. All relevant antiderivatives in this paper are derived from improper integrals with format $\int_1^\infty f(n)dn$ based on Eqs. (6), (17) & (19). Example for Eq. (6), involved improper integrals are from $\int_1^\infty (2n)^{-\sigma}\sqrt{2}\sin(t\ln(2n) + \frac{3}{4}\pi) - \int_1^\infty (2n-1)^{-\sigma}\sqrt{2}\sin(t\ln(2n-1) + \frac{3}{4}\pi) = 0$. These improper integrals are seen to involve [periodic] sine function between limits 1 and ∞. Each improper integral can be validly expanded as $\int_{n=1}^{n=2} f(n)dn + \int_{n=2}^{n=3} f(n)dn + \int_{n=3}^{n=4} f(n)dn +...+ \int_{n=\infty-1}^{n=\infty} f(n)dn$ which, for all sufficiently large t as t$\longrightarrow \infty$, will manifest *divergence by oscillation* (viz. for all sufficiently large t as t$\longrightarrow \infty$, this cummulative total will not diverge in a particular direction to a solitary well-defined limit value such as $\sin \pi/2 = 1$ or less well-defined limit value such as $+\infty$). Note that all types of Dirichlet Sigma-Power Laws are parametric equations (and inequations) where t is the parameter.

With steps of manual integration shown using indefinite integrals [for simplicity], we solve definite integral based on numerator portion of R1 with $(2n)$ parameter in Eq. (6):

$$\int_1^\infty \frac{2^{\frac{1}{2}-\sigma}\sin\left(t\ln(2n) + \frac{3\pi}{4}\right)}{n^\sigma} dn = \int_1^\infty -\frac{\sin(t\ln(2n)) - \cos(t\ln(2n))}{2^\sigma n^\sigma} dn.$$ We deduce most other important integrals to be "variations" of this particular integral containing (i) deletion of $(2n)^{-\sigma}$, $\sqrt{2}$ or $\frac{3}{4}\pi$ terms, and/or (ii) interchange of sine and cosine function. We check all derived antiderivatives to be correct using computer algebra system Maxima.

Simplifying and applying linearity, we obtain $2^{\frac{1}{2}-\sigma}\int \frac{\sin\left(t\ln(2n) + \frac{3\pi}{4}\right)}{n^\sigma} dn$.

Now solving $\int \frac{\sin\left(t\ln(2n) + \frac{3\pi}{4}\right)}{n^\sigma} dn$. Substitute $u = t\ln(2n) + \frac{3\pi}{4}$ $\longrightarrow dn = \frac{n}{t} du$, use $n^{1-\sigma} =$

$$e^{\frac{(1-\sigma)\left(u-t\ln(2)-\frac{3\pi}{4}\right)}{t}} = \frac{e^{\frac{(\sigma-1)(4t\ln(2)+3\pi)}{4t}}}{t} \int e^{\frac{(1-\sigma)u}{t}} \sin(u)\, du.$$

Now solving $\int e^{\frac{(1-\sigma)u}{t}} \sin(u)\, du$. We integrate by parts twice in a row: $\int \mathbf{fg'} = \mathbf{fg} - \int \mathbf{f'g}$.

First time: f = $\sin(u)$, $g' = e^{\frac{(1-\sigma)u}{t}}$

Then $\mathbf{f'} = \boxed{\cos(u)}, \mathbf{g} = \boxed{\dfrac{(1-\sigma)te^{\frac{(1-\sigma)u}{t}}}{\sigma^2 - 2\sigma + 1}}$:

$$= \frac{(1-\sigma)te^{\frac{(1-\sigma)u}{t}}\sin(u)}{\sigma^2 - 2\sigma + 1} - \int \frac{(1-\sigma)te^{\frac{(1-\sigma)u}{t}}\cos(u)}{\sigma^2 - 2\sigma + 1}\,du$$

Second time: $\mathbf{f} = \boxed{\cos(u)}, \mathbf{g'} = \boxed{\dfrac{(1-\sigma)te^{\frac{(1-\sigma)u}{t}}}{\sigma^2 - 2\sigma + 1}}$

Then $\mathbf{f'} = -\sin(u), \mathbf{g} = \dfrac{t^2 e^{\frac{(1-\sigma)u}{t}}}{\sigma^2 - 2\sigma + 1}$:

$$= \frac{(1-\sigma)te^{\frac{(1-\sigma)u}{t}}\sin(u)}{\sigma^2 - 2\sigma + 1} - \left(\frac{t^2 e^{\frac{(1-\sigma)u}{t}}\cos(u)}{\sigma^2 - 2\sigma + 1} - \int -\frac{t^2 e^{\frac{(1-\sigma)u}{t}}\sin(u)}{\sigma^2 - 2\sigma + 1}\,du \right)$$

Apply linearity:

$$= \frac{(1-\sigma)te^{\frac{(1-\sigma)u}{t}}\sin(u)}{\sigma^2 - 2\sigma + 1} - \left(\frac{t^2 e^{\frac{(1-\sigma)u}{t}}\cos(u)}{\sigma^2 - 2\sigma + 1} + \frac{t^2}{\sigma^2 - 2\sigma + 1}\int e^{\frac{(1-\sigma)u}{t}}\sin(u)\,du \right)$$

As integral $\int e^{\frac{(1-\sigma)u}{t}}\sin(u)\,du$ appears again on Right Hand Side, we solve for it:

$$= \frac{\frac{(1-\sigma)e^{\frac{(1-\sigma)u}{t}}\sin(u)}{t} - e^{\frac{(1-\sigma)u}{t}}\cos(u)}{\frac{\sigma^2 - 2\sigma + 1}{t^2} + 1}$$

Plug in solved integrals: $\dfrac{e^{\frac{(\sigma-1)(4t\ln(2)+3\pi)}{4t}}}{t}\int e^{\frac{(1-\sigma)u}{t}}\sin(u)\,du$

$$= \frac{e^{\frac{(\sigma-1)(4t\ln(2)+3\pi)}{4t}} \left(\frac{(1-\sigma)e^{\frac{(1-\sigma)u}{t}}\sin(u)}{t} - e^{\frac{(1-\sigma)u}{t}}\cos(u) \right)}{\left(\frac{\sigma^2 - 2\sigma + 1}{t^2} + 1 \right)t}$$

Undo substitution $u = t\ln(2n) + \frac{3\pi}{4}$ and simplifying:

$$= \frac{e^{\frac{(\sigma-1)(4t\ln(2)+3\pi)}{4t}} \left(\frac{(1-\sigma)e^{\frac{(1-\sigma)\left(t\ln(2n)+\frac{3\pi}{4}\right)}{t}}\sin\left(t\ln(2n)+\frac{3\pi}{4}\right)}{t} - e^{\frac{(1-\sigma)\left(t\ln(2n)+\frac{3\pi}{4}\right)}{t}}\cos\left(t\ln(2n)+\frac{3\pi}{4}\right) \right)}{\left(\frac{\sigma^2 - 2\sigma + 1}{t^2} + 1 \right)t}$$

Plug in solved integrals: $2^{\frac{1}{2}-\sigma}\int \dfrac{\sin\left(t\ln(2n)+\frac{3\pi}{4}\right)}{n^\sigma}\,dn$

$$= \frac{2^{\frac{1}{2}-\sigma} e^{\frac{(\sigma-1)(4t\ln(2)+3\pi)}{4t}} \left(\frac{(1-\sigma)e^{\frac{(1-\sigma)\left(t\ln(2n)+\frac{3\pi}{4}\right)}{t}}\sin\left(t\ln(2n)+\frac{3\pi}{4}\right)}{t} - e^{\frac{(1-\sigma)\left(t\ln(2n)+\frac{3\pi}{4}\right)}{t}}\cos\left(t\ln(2n)+\frac{3\pi}{4}\right) \right)}{\left(\frac{\sigma^2 - 2\sigma + 1}{t^2} + 1 \right)t}$$

By rewriting and simplifying, $\displaystyle\int_1^\infty \frac{2^{\frac{1}{2}-\sigma}\sin\left(t\ln(2n)+\frac{3\pi}{4}\right)}{n^\sigma}\,dn$ is finally solved as

$$\left[\frac{(2n)^{1-\sigma}\left((t+\sigma-1)\sin(t\ln(2n)) + (t-\sigma+1)\cos(t\ln(2n))\right)}{2\left(t^2 + (\sigma-1)^2\right)} + C \right]_1^\infty \tag{7}$$

For denominator portion of R1 with $(2n-1)$ parameter in Eq. (6), Eq. (7) equates to

$$\left[\frac{(2n-1)^{1-\sigma}\left((t+\sigma-1)\sin\left(t\ln\left(2n-1\right)\right)+(t-\sigma+1)\cos\left(t\ln\left(2n-1\right)\right)\right)}{2\left(t^2+(\sigma-1)^2\right)}+C\right]_1^\infty \quad (8)$$

Dirichlet Sigma-Power Law as equation derived from Eq. (6) is given by:

$$\frac{1}{2(t^2+(\sigma-1)^2)} \cdot [(2n)^{1-\sigma}\left((t+\sigma-1)\sin\left(t\ln\left(2n\right)\right)+(t-\sigma+1)\cos\left(t\ln\left(2n\right)\right)\right)-$$

$$(2n-1)^{1-\sigma}\left((t+\sigma-1)\sin\left(t\ln\left(2n-1\right)\right)+(t-\sigma+1)\cos\left(t\ln\left(2n-1\right)\right)\right)]_1^\infty = 0 \quad (9)$$

Apply Ratio Study to Eq. (6) – see Appendix B. This involves [intentional] incorrect but "balanced" rearrangement of terms in Eq. (6) giving rise to Eq. (10) which is a non-Hybrid integer sequence inequation. Left-hand side contains 'cyclical' sine function in first term (Ratio R1) and 'non-cyclical' power function in second term (Ratio R2).

$$\frac{\sum_{n=1}^\infty \sqrt{2}\sin(t\ln(2n)+\frac{3}{4}\pi)}{\sum_{n=1}^\infty \sqrt{2}\sin(t\ln(2n-1)+\frac{3}{4}\pi)} - \frac{\sum_{n=1}^\infty (2n)^\sigma}{\sum_{n=1}^\infty (2n-1)^\sigma} \neq 0 \quad (10)$$

Apply Riemann integral to selected parts of Eq. (10) without depicting steps of calculation:

$$\int_1^\infty \sqrt{2}\sin\left(t\ln\left(2n\right)+\frac{3\pi}{4}\right)dn =$$
$$\left[\frac{(2n)\left((t-1)\sin\left(t\ln\left(2n\right)\right)+(t+1)\cos\left(t\ln\left(2n\right)\right)\right)}{2(t^2+1)}+C\right]_1^\infty$$

and $\int_1^\infty \sqrt{2}\sin\left(t\ln\left(2n-1\right)+\frac{3\pi}{4}\right)dn =$
$$\left[\frac{(2n-1)\left((t-1)\sin\left(t\ln\left(2n-1\right)\right)+(t+1)\cos\left(t\ln\left(2n-1\right)\right)\right)}{2(t^2+1)}+C\right]_1^\infty$$

$$\int_1^\infty (2n)^\sigma dn = \left[\frac{(2n)^{\sigma+1}}{2(\sigma+1)}+C\right]_1^\infty \text{ and } \int_1^\infty (2n-1)^\sigma dn = \left[\frac{(2n-1)^{\sigma+1}}{2(\sigma+1)}+C\right]_1^\infty$$

Dirichlet Sigma-Power Law as inequation derived from Eq. (10) is given by:

$$\left[\frac{(2n)\left((t-1)\sin\left(t\ln\left(2n\right)\right)+(t+1)\cos\left(t\ln\left(2n\right)\right)\right)}{(2n-1)\left((t-1)\sin\left(t\ln\left(2n-1\right)\right)+(t+1)\cos\left(t\ln\left(2n-1\right)\right)\right)} - \frac{(2n)^{\sigma+1}}{(2n-1)^{\sigma+1}}\right]_1^\infty$$
$$\neq 0 \quad (11)$$

Intended derivation of Dirichlet Sigma-Power Law as equation and inequation have been successful. *The proof is now complete for Proposition 3.2*□.

Proposition 3.3. Exact Dimensional analysis homogeneity at $\sigma = \frac{1}{2}$ in Dirichlet Sigma-Power Law as equation and inequation is (respectively) indicated by $\sum$(all fractional exponents) = whole number '1' and '3'.

Proof. Preliminary discussion on using three types of symmetry for a given function: (1) symmetry about the vertical y-axis ["function is even"] e.g. cosine, arccos (2) symmetry about the origin ["function is odd"] e.g. sine, arcsin, tangent, arctan and (3) in all other cases ["function is neither even nor odd"]. Even function has its Cummulative Total areas symmetrical about the vertical axis and odd function has its Cummulative Total areas symmetrical about the origin (with conservation or preservation of areas derived from [opposite side] numerical "net area value" always equal to zero in both cases).

We classify our antiderivatives below using these functions with their basic properties such as sum [or difference] of two even (odd) functions is even (odd); sum [or difference] of an even and odd function is neither even nor odd (unless one of the functions is equal to zero over the given domain); product [or quotient] of two even or odd functions is an even function; and product [or quotient] of an even function and an odd function is an odd function. We will shortly see that only Dirichlet Sigma-Power Laws as equation and inequation pertaining to calculations intended for Gram[x=0,y=0] points will uniquely manifest "functions that are neither even nor odd".

Dirichlet Sigma-Power Law as equation for $\sigma = \frac{1}{2}$ value is given by:

$$\frac{1}{2t^2+\frac{1}{2}} \cdot [(2n)^{\frac{1}{2}} \left((t-\tfrac{1}{2})\sin(t\ln(2n)) + (t+\tfrac{1}{2})\cos(t\ln(2n))\right) -$$

$$(2n-1)^{\frac{1}{2}}\left((t-\tfrac{1}{2})\sin(t\ln(2n-1)) + (t+\tfrac{1}{2})\cos(t\ln(2n-1))\right)]_1^\infty = 0 \qquad (12)$$

Respectively evaluation of definite integrals Eq. (12), Eq. (24) and Eq. (26) using limit as $n \to +\infty$ for $0 < t < +\infty$ enable countless computations resulting in t values for CIS of nontrivial zeros, Gram[y=0] and Gram[x=0] points. Note that larger n values used for computations will correspond to increasing accuracy of these entities (which are all transcendental numbers). We evaluate Eq. (12) to obtain its expanded antiderivative:

$$\frac{1}{2t^2+\frac{1}{2}} \cdot [(2\infty)^{\frac{1}{2}}\left((t-\tfrac{1}{2})\sin(t\ln(2\infty)) + (t+\tfrac{1}{2})\cos(t\ln(2\infty))\right) -$$

$$(2\infty-1)^{\frac{1}{2}}\left((t-\tfrac{1}{2})\sin(t\ln(2\infty-1)) + (t+\tfrac{1}{2})\cos(t\ln(2\infty-1))\right)$$

$$-(2)^{\frac{1}{2}}\left((t-\tfrac{1}{2})\sin(t\ln(2)) + (t+\tfrac{1}{2})\cos(t\ln(2))\right) +$$

$$(1)^{\frac{1}{2}}\left((t-\tfrac{1}{2})\sin(t\ln(1)) + (t+\tfrac{1}{2})\cos(t\ln(1))\right)] = 0$$

$$\frac{1}{2t^2+\frac{1}{2}} \cdot [(2\infty)^{\frac{1}{2}}\left((t-\tfrac{1}{2})\sin(t\ln(2\infty)) + (t+\tfrac{1}{2})\cos(t\ln(2\infty))\right) -$$

$$(2\infty-1)^{\frac{1}{2}}\left((t-\tfrac{1}{2})\sin(t\ln(2\infty-1)) + (t+\tfrac{1}{2})\cos(t\ln(2\infty-1))\right)$$

$$-(2)^{\frac{1}{2}}\left((t-\tfrac{1}{2})\sin(t\ln(2)) + (t+\tfrac{1}{2})\cos(t\ln(2))\right) + (t+\tfrac{1}{2})] = 0$$

We are interested in the last two terms [equivalent to substituting in n = 1]

$$\frac{1}{2t^2+\frac{1}{2}} \cdot [-(2)^{\frac{1}{2}}\left((t-\tfrac{1}{2})\sin(t\ln(2)) + (t+\tfrac{1}{2})\cos(t\ln(2))\right) + (t+\tfrac{1}{2})]$$

Equivalent evaluation on Eq. (12) to obtain its expanded antiderivative depicted as linear combination of sine and cosine waves: $a\sin x + b\cos x = c\sin(x + \varphi)$ with $c = \sqrt{a^2+b^2}$ and $\varphi = \mathrm{atan2}(b,a) = \tan^{-1}(\tfrac{b}{a})$ for $a>0$:

$$\frac{1}{2t^2+\frac{1}{2}} \cdot [((2\infty)(2t^2+\tfrac{1}{2}))^{\frac{1}{2}} \sin\left((t\ln 2\infty) + \tan^{-1}(\tfrac{t+\frac{1}{2}}{t-\frac{1}{2}})\right)$$

$$-((2\infty-1)(2t^2+\tfrac{1}{2}))^{\frac{1}{2}} \sin\left((t\ln 2\infty - 1) + \tan^{-1}(\tfrac{t+\frac{1}{2}}{t-\frac{1}{2}})\right) - ((2)(2t^2+\tfrac{1}{2}))^{\frac{1}{2}}$$

$$\sin\left((t\ln 2) + \tan^{-1}(\tfrac{t+\frac{1}{2}}{t-\frac{1}{2}})\right) + (2t^2+\tfrac{1}{2})^{\frac{1}{2}} \sin\left(\tan^{-1}(\tfrac{t+\frac{1}{2}}{t-\frac{1}{2}})\right)] = 0$$

We look at the last two terms

$$\frac{1}{2t^2+\frac{1}{2}} \cdot [-((2)(2t^2+\frac{1}{2}))^{\frac{1}{2}} \sin\left((t\ln 2) + \tan^{-1}(\frac{t+\frac{1}{2}}{t-\frac{1}{2}})\right) + (2t^2+\frac{1}{2})^{\frac{1}{2}} \sin\left(\tan^{-1}(\frac{t+\frac{1}{2}}{t-\frac{1}{2}})\right)]$$

Relevant t values for all nontrivial zeros at $\sigma = \frac{1}{2}$ and n = 1 [and combined n = 1, 2 and 3]

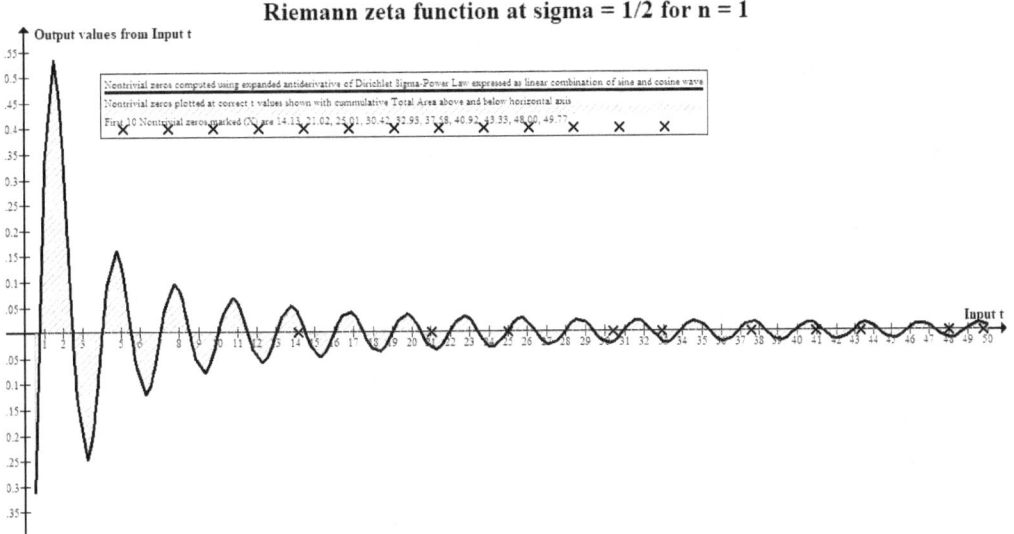

FIGURE 5. Nontrivial zeros [first 10 plotted] at $\sigma = \frac{1}{2}$ for n = 1 using Dirichlet Sigma-Power Law.

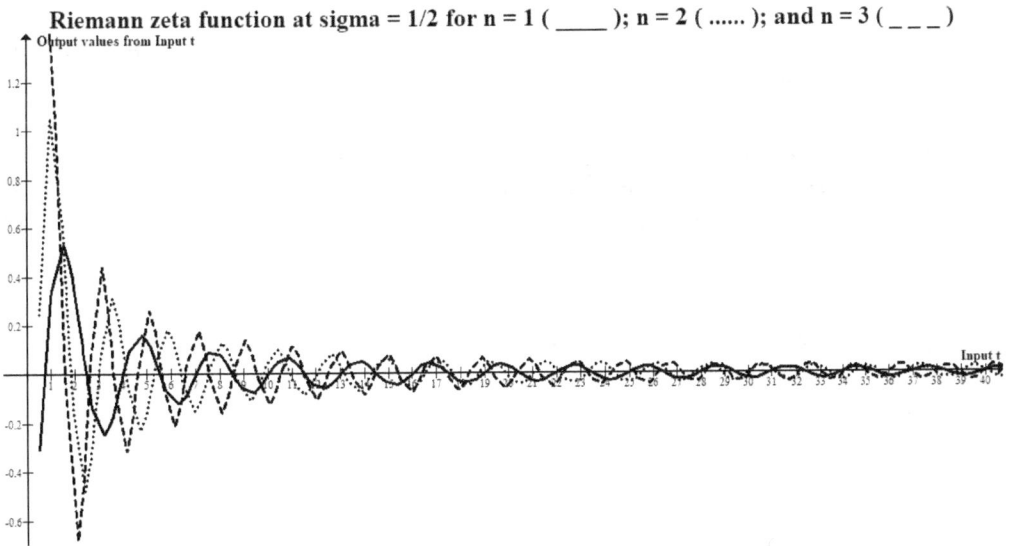

FIGURE 6. Nontrivial zeros at $\sigma = \frac{1}{2}$ for n = 1, 2 and 3 using Dirichlet Sigma-Power Law.

plotted against the expanded antiderivative depicted as linear combination of sine and cosine waves is shown in Figure 5 [and in Figure 6]. The phenomenon of "monotonously decreasing height waves of varying amplitude with different n values" comply with functions that are neither even nor odd. As mentioned previously, all improper integrals are seen to involve [periodic] sine function between limits 1 and ∞. Then from Figure 6 for nontrivial zeros, we can geometrically visualize that fully computing (for instance) $\int_{n=1}^{n=2} f(n)dn$ and $\int_{n=2}^{n=3} f(n)dn$ will result in respective antiderivatives that still involve sine functions [of varying frequency and amplitude]. Identical arguments can be extended to Gram[y=0] and Gram[x=0] points.

Dirichlet Sigma-Power Law as inequation for $\sigma = \frac{1}{2}$ value is given by:

$$\left[\frac{(2n)\left((t-1)\sin(t\ln(2n)) + (t+1)\cos(t\ln(2n))\right)}{(2n-1)\left((t-1)\sin(t\ln(2n-1)) + (t+1)\cos(t\ln(2n-1))\right)} - \frac{(2n)^{\frac{3}{2}}}{(2n-1)^{\frac{3}{2}}}\right]_1^\infty \neq 0 \qquad (13)$$

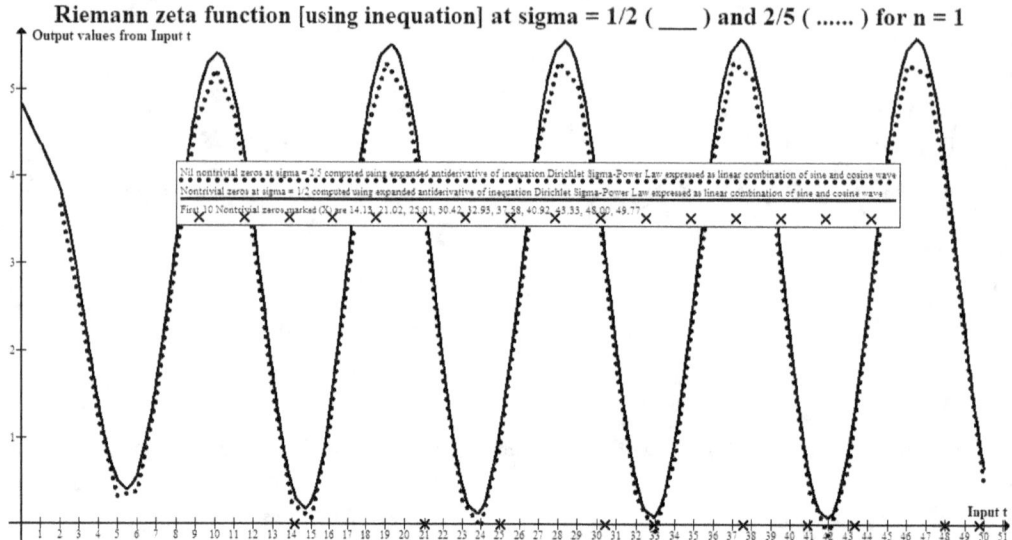

FIGURE 7. Nontrivial zeros [first 10 plotted] present at $\sigma = \frac{1}{2}$ and nil nontrivial zeros present at $\sigma = \frac{2}{5}$ with both calculated for n = 1 using **inequation** Dirichlet Sigma-Power Law.

Without depicting parallel steps of calculation, the equivalent **inequations** Dirichlet Sigma-Power Law for last two terms [expanded antiderivative depicted as linear combination of sine and cosine waves] for nontrivial zeros [first 10 plotted] at $\sigma = \frac{1}{2}$ and $\sigma = \frac{2}{5}$ when n = 1 is, respectively and sequentially, given below as:

$$-\frac{(2)\left((2t^2+2)^{\frac{1}{2}}\sin((t\ln 2) + \tan^{-1}(\frac{t+1}{t-1}))\right)}{(1)\left((2t^2+2)^{\frac{1}{2}}\sin((t\ln 1) + \tan^{-1}(\frac{t+1}{t-1}))\right)} + \frac{(2)^{\frac{3}{2}}}{(1)^{\frac{3}{2}}}$$

$$-\frac{(2)\left((2t^2+2)^{\frac{1}{2}}\sin((t\ln 2) + \tan^{-1}(\frac{t+1}{t-1}))\right)}{(1)\left((2t^2+2)^{\frac{1}{2}}\sin((t\ln 1) + \tan^{-1}(\frac{t+1}{t-1}))\right)} + \frac{(2)^{\frac{7}{5}}}{(1)^{\frac{7}{5}}}$$

Figure 7 depicted using the two inequations will manifest phenomenon of "constant height waves of varying amplitude with different n values" comply with functions that are neither even nor odd occurring at both $\sigma = \frac{1}{2}$ and $\sigma \neq \frac{1}{2}$.

$\sum$(all fractional exponents) as $2(1-\sigma)$ = whole number '1' for Eq. (12) and $2(\sigma+1)$ = whole number '3' for Eq. (13). These findings signify presence of complete set nontrivial zeros for Eq. (12) and Eq. (13). *The proof is now complete for Proposition 3.3*□.

Corollary 3.4. Inexact Dimensional analysis homogeneity at $\sigma \neq \frac{1}{2}$ [illustrated using $\sigma = \frac{2}{5}$] in Dirichlet Sigma-Power Law as equation and inequation is (respectively) indicated by $\sum$(all fractional exponents) = fractional number '$\neq 1$' and '$\neq 3$'.

Proof. Dirichlet Sigma-Power Law as equation for $\sigma = \frac{2}{5}$ value is given by:

$$\frac{1}{2t^2 + \frac{18}{25}} \cdot [(2n)^{\frac{3}{5}}\left((t - \frac{3}{5})\sin(t\ln(2n)) + (t + \frac{3}{5})\cos(t\ln(2n))\right) -$$

$$(2n - 1)^{\frac{3}{5}}\left((t - \frac{3}{5})\sin(t\ln(2n-1)) + (t + \frac{3}{5})\cos(t\ln(2n-1))\right)]_1^\infty = 0 \quad (14)$$

We evaluate Eq. (14) to obtain its expanded antiderivative:

$$\frac{1}{2t^2 + \frac{18}{25}} \cdot [(2\infty)^{\frac{3}{5}}\left((t - \frac{3}{5})\sin(t\ln(2\infty)) + (t + \frac{3}{5})\cos(t\ln(2\infty))\right) -$$

$$(2\infty - 1)^{\frac{3}{5}}\left((t - \frac{3}{5})\sin(t\ln(2\infty - 1)) + (t + \frac{3}{5})\cos(t\ln(2\infty - 1))\right)$$

$$-(2)^{\frac{3}{5}}\left((t - \frac{3}{5})\sin(t\ln(2)) + (t + \frac{3}{5})\cos(t\ln(2))\right) +$$

$$(1)^{\frac{3}{5}}\left((t - \frac{3}{5})\sin(t\ln(1)) + (t + \frac{3}{5})\cos(t\ln(1))\right)] = 0$$

$$\frac{1}{2t^2 + \frac{18}{25}} \cdot [(2\infty)^{\frac{3}{5}}\left((t - \frac{3}{5})\sin(t\ln(2\infty)) + (t + \frac{3}{5})\cos(t\ln(2\infty))\right) -$$

$$(2\infty - 1)^{\frac{3}{5}}\left((t - \frac{3}{5})\sin(t\ln(2\infty - 1)) + (t + \frac{3}{5})\cos(t\ln(2\infty - 1))\right)$$

$$-(2)^{\frac{3}{5}}\left((t - \frac{3}{5})\sin(t\ln(2)) + (t + \frac{3}{5})\cos(t\ln(2))\right)$$

We are interested in the last two terms [equivalent to substituting in n = 1]

$$\frac{1}{2t^2 + \frac{18}{25}} \cdot [-(2)^{\frac{3}{5}}\left((t - \frac{3}{5})\sin(t\ln(2)) + (t + \frac{3}{5})\cos(t\ln(2))\right) + (t + \frac{3}{5})]$$

Equivalent valuation on Eq. (14) to obtain its expanded antiderivative depicted as linear combination of sine and cosine waves: $a\sin x + b\cos x = c\sin(x + \varphi)$ with $c = \sqrt{a^2 + b^2}$ and $\varphi = \text{atan2}(b, a)$

$= \tan^{-1}(\frac{b}{a})$ for a>0: $\frac{1}{2t^2 + \frac{18}{25}} \cdot [((2\infty)^{\frac{3}{5}}(2t^2 + \frac{18}{25}))^{\frac{1}{2}} \sin\left((t\ln 2\infty) + \tan^{-1}(\frac{t + \frac{3}{5}}{t - \frac{3}{5}})\right)$

$-((2\infty - 1)^{\frac{3}{5}}(2t^2 + \frac{18}{25}))^{\frac{1}{2}} \sin\left((t\ln 2\infty - 1) + \tan^{-1}(\frac{t + \frac{3}{5}}{t - \frac{3}{5}})\right) - ((2)^{\frac{3}{5}}(2t^2 + \frac{18}{25}))^{\frac{1}{2}} \sin\left((t\ln 2) + \tan^{-1}(\frac{t + \frac{3}{5}}{t - \frac{3}{5}})\right)$

$+(2t^2 + \frac{18}{25})^{\frac{1}{2}} \sin\left(\tan^{-1}(\frac{t + \frac{3}{5}}{t - \frac{3}{5}})\right)] = 0$

We look at the last two terms

$$\frac{1}{2t^2 + \frac{18}{25}} \cdot [-(2^{\frac{3}{5}}(2t^2 + \frac{18}{25}))^{\frac{1}{2}} \sin\left((t\ln 2) + \tan^{-1}(\frac{t + \frac{3}{5}}{t - \frac{3}{5}})\right) + (2t^2 + \frac{18}{25})^{\frac{1}{2}} \sin\left(\tan^{-1}(\frac{t + \frac{3}{5}}{t - \frac{3}{5}})\right)]$$

Relevant t values (for non-existent nontrivial zeros) at $\sigma = \frac{2}{5}$ [and $\sigma = \frac{3}{5}$] for n = 1 plotted against the expanded antiderivative depicted as linear combination of sine and cosine waves is shown in Figure 8 [and in Figure 9]. The phenomenon of "monotonously decreasing height waves of varying amplitude with different n values" comply with functions that are neither even nor odd occurring at $\sigma \neq \frac{1}{2}$.

Dirichlet Sigma-Power Law as inequation for $\sigma = \frac{2}{5}$ value [previously depicted in Figure 7 above from Proposition 3.3] is given by:

$$\left[\frac{(2n)\left((t-1)\sin(t\ln(2n)) + (t+1)\cos(t\ln(2n))\right)}{(2n-1)\left((t-1)\sin(t\ln(2n-1)) + (t+1)\cos(t\ln(2n-1))\right)} - \frac{(2n)^{\frac{7}{5}}}{(2n-1)^{\frac{7}{5}}}\right]_1^\infty \neq 0 \quad (15)$$

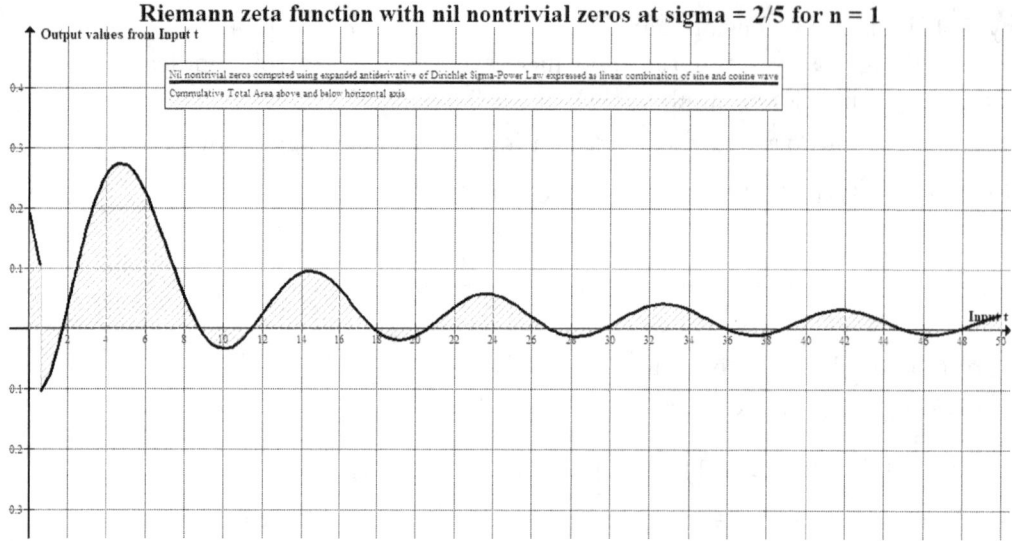

FIGURE 8. Nil nontrivial zeros present at $\sigma = \frac{2}{5}$ for n = 1 using Dirichlet Sigma-Power Law.

$\sum$(all fractional exponents) as $2(1-\sigma)$ = fractional number '$\neq 1$' for Eq. (14) and $2(\sigma+1)$ = fractional number '$\neq 3$' for Eq. (15). These findings signify absence of complete set nontrivial zeros for Eq. (14) and Eq. (15). *The proof is now complete for Corollary 3.4*□.

4. Rigorous Proof for Riemann hypothesis summarized as Theorem Riemann I – IV

$\zeta(s) = \frac{1}{s-1} + \frac{1}{2} + 2\int_0^\infty \frac{\sin(s \arctan t)}{(1+t^2)^{\frac{s}{2}}(e^{2\pi t}-1)} dt$ is integral relation (cf. Abel-Plana summation formula [3][4]) for all s∈ℂ and s≠1. This integral is insufficient for our purpose as it involves integration with respect to t [instead of n] for $\zeta(s)$ [instead of $\eta(s)$]. Relatively elementary proof for Riemann hypothesis is summarized by Theorem Riemann I – IV. One could obtain this proof with only using Dirichlet Sigma-Power Law [solely] as equation. For completeness and clarification of this proof, we supply following important mathematical arguments.

For $0 < \sigma < 1$, then $0 < 2(1-\sigma) < 2$. The only whole number between 0 and 2 is '1' which coincide with $\sigma = \frac{1}{2}$. When $0 < \sigma < \frac{1}{2}$ and $\frac{1}{2} < \sigma < 1$, then $0 < 2(1-\sigma) < 1$ and $1 < 2(1-\sigma) < 2$.

For $0 < \sigma < 1$, $2 < 2(\sigma+1) < 4$. The only whole number between 2 and 4 is '3' which coincide with $\sigma = \frac{1}{2}$. When $0 < \sigma < \frac{1}{2}$ and $\frac{1}{2} < \sigma < 1$, then $2 < 2(\sigma+1) < 3$ and $3 < 2(\sigma+1) < 4$.

Legend: **R** = all real numbers. For $0 < \sigma < 1$, σ consist of $0 < \mathbf{R} < 1$. For $0 < 2(1-\sigma) < 2$ and $2 < 2(\sigma+1) < 4$, $2(1-\sigma)$ and $2(\sigma+1)$ must (respectively) consist of $0 < \mathbf{R} < 2$ and $2 < \mathbf{R} < 4$. An important caveat is that previously used phrases such as "fractional exponent σ" and "$\sum$(all fractional exponents) = whole number '1' [or '3'] and fractional number '$\neq 1$' [or '$\neq 3$']", although not incorrect *per se*, should respectively be replaced by "real number exponent σ" and "$\sum$(all real number exponents) = whole number '1' [or '3'] and real number '$\neq 1$' [or '$\neq 3$'][5]" for complete accurracy. We apply this caveat to Theorem Riemann I – IV.

Footnote 5: As whole numbers ⊂ real numbers, one could also depict this phrase as "$\sum$(all

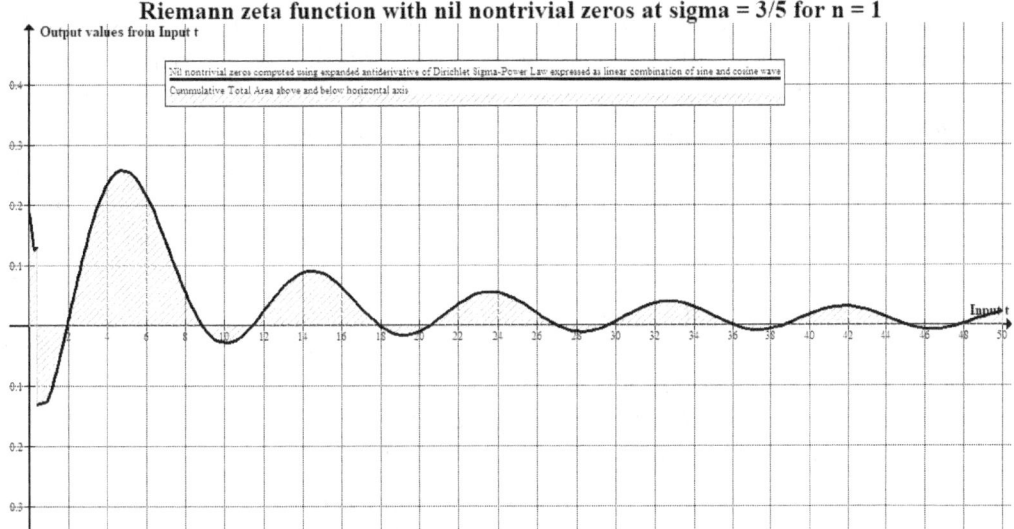

FIGURE 9. Nil nontrivial zeros present at $\sigma = \frac{3}{5}$ for n = 1 using Dirichlet Sigma-Power Law with last two terms of expanded antiderivative depicted as linear combination of sine and cosine waves given by

$$\frac{1}{2t^2 + \frac{8}{25}} \cdot [-(2^{\frac{2}{5}}(2t^2 + \frac{8}{25}))^{\frac{1}{2}} \sin\left((t\ln 2) + \tan^{-1}(\frac{t+\frac{2}{5}}{t-\frac{2}{5}})\right) + (2t^2 + \frac{8}{25})^{\frac{1}{2}} \sin\left(\tan^{-1}(\frac{t+\frac{2}{5}}{t-\frac{2}{5}})\right)].$$

real number exponents) = real number '1' [or '3'] and real number '≠1' [or '≠3']".

Theorem Riemann I. Derived from *proxy* Dirichlet eta function, "simplified" Dirichlet eta function will exclusively contain *de novo* property for actual location [but not actual positions] of all nontrivial zeros.

Proof. The phrase "actual location [but not actual positions] of all nontrivial zeros" can be validly shortened to "actual location of all nontrivial zeros" as used in Theorem Riemann II, III and IV. *The proof for Theorem Riemann I is now complete as it successfully incorporates proof for Lemma 3.1*□.

Theorem Riemann II. Dirichlet Sigma-Power Law [in continuous (integral) format] as equation and inequation which are both derived from "simplified" Dirichlet eta function [in discrete (summation) format] will exclusively manifest exact DA homogeneity in equation and inequation only when real number exponent $\sigma = \frac{1}{2}$.

Proof. *The proof for Theorem Riemann II is now complete as it successfully incorporates proofs from Proposition 3.2 on derivation for equation and inequation of Dirichlet Sigma-Power Law [with both containing de novo property for "actual location of all nontrivial zeros"] and Proposition 3.3 on manifestation of exact DA homogeneity in Dirichlet Sigma-Power Law as equation and inequation when real number exponent $\sigma = \frac{1}{2}$*□.

Theorem Riemann III. Real number exponent $\sigma = \frac{1}{2}$ in Dirichlet Sigma-Power Law as equation and inequation satisfying exact DA homogeneity is identical to σ variable in Riemann hypothesis which propose σ to also have exclusive value of $\frac{1}{2}$ (representing critical line) for "actual location of all nontrivial zeros", thus fully supporting Riemann hypothesis to be true with further clarification by Theorem Riemann IV.

Proof. Since $s = \sigma \pm \imath t$, complete set of nontrivial zeros which is defined by $\eta(s) = 0$ is exclusively associated with one (and only one) particular $\eta(\sigma \pm \imath t) = 0$ value solution, and by default one (and only one) particular σ [conjecturally] $= \frac{1}{2}$ value solution. When performing exact DA homogeneity on Dirichlet Sigma-Power Law as equation and inequation [with both containing *de novo* property for "actual location of all nontrivial zeros"], the phrase "If real number exponent σ has exclusively $\frac{1}{2}$ value, only then will exact DA homogeneity be satisfied" implies one (and only one) possible mathematical solution. Theorem Riemann III reflect Theorem Riemann II on presence of exact DA homogeneity for $\sigma = \frac{1}{2}$ in Dirichlet Sigma-Power Law as equation and inequation. This Law has identical σ variable as that referred to by Riemann hypothesis [whereby σ here uniquely refer to critical line]. *The proof for Theorem Riemann III is now complete as it independently refers to simultaneous association of confirmed (i) solitary $\sigma = \frac{1}{2}$ value in Dirichlet Sigma-Power Law as equation and inequation satisfying exact DA homogeneity and (ii) critical line defined by solitary $\sigma = \frac{1}{2}$ value being the "actual location [but with no request to determine actual positions]" of all nontrivial zeros as proposed in original Riemann hypothesis*□.

Theorem Riemann IV. Condition 1. All $\sigma \neq \frac{1}{2}$ values (non-critical lines), viz. $0 < \sigma < \frac{1}{2}$ and $\frac{1}{2} < \sigma < 1$ values, exclusively does not contain "actual location of all nontrivial zeros" [manifesting *de novo* inexact DA homogeneity in equation and inequation], together with Condition 2. One (& only one) $\sigma = \frac{1}{2}$ value (critical line) exclusively contains "actual location of all nontrivial zeros" [manifesting *de novo* exact DA homogeneity in equation and inequation], confirm Riemann hypothesis to be true when these two mutually inclusive conditions are met.

Proof. Condition 2 Theorem Riemann IV simply reflect proof from Theorem Riemann III [incorporating Proposition 3.3] for "actual location of all nontrivial zeros" exclusively on critical line manifesting *de novo* exact DA homogeneity $\sum$(all real number exponents) = whole number '1' for equation [or '3' for inequation]. *The proof for Condition 2 Theorem Riemann IV is now complete*□. Corollary 3.4 confirms *de novo* inexact DA homogeneity manifested as $\sum$(all real number exponents) = real number '$\neq 1$' for equation [or '$\neq 3$' for inequation] by all $\sigma \neq \frac{1}{2}$ values (non-critical lines) that are exclusively not associated with "actual location of all nontrivial zeros". Applying inclusion-exclusion principle: Exclusive presence of nontrivial zeros on critical line for Condition 2 Theorem Riemann IV implies exclusive absence of nontrivial zeros on non-critical lines for Condition 1 Theorem Riemann IV. *The proof for Condition 1 Theorem Riemann IV is now complete*□.

We logically deduce that explicit mathematical explanation why presence and absence of nontrivial zeros[6] should (respectively) coincide precisely with $\sigma = \frac{1}{2}$ and $\sigma \neq \frac{1}{2}$ [literally the Completely Predictable meta-properties ('overall' *complex properties*)] will require "complex" mathematical arguments. Attempting to provide explicit mathematical explanation with "simple" mathematical arguments would intuitively mean nontrivial zeros have to be (incorrectly and impossibly) treated as Completely Predictable entities.

Footnote 6: Completely Predictable meta-properties for Gram and virtual Gram points equating to "Presence of Gram[y=0] and Gram[x=0] points, and virtual Gram[y=0] and virtual Gram[x=0] points (respectively) coincide precisely with $\sigma = \frac{1}{2}$, and $\sigma \neq \frac{1}{2}$".

5. Prerequisite lemma, corollary and propositions for Gram[x=0] and Gram[y=0] points conjectures

For Gram[y=0] & Gram[x=0] points (and corresponding virtual Gram[y=0] & virtual Gram[x=0] points with totally different values), we apply a parallel procedure carried out on nontrivial zeros but only depict abbreviated treatments and discussions. We supply geometrical manifestations and related commentaries for equivalent "last two terms" at n = 1 for each entity using expanded antiderivative (at $\sigma = \frac{1}{2}$ & $\frac{2}{5}$) depicted as linear combination of sine and cosine waves: $a\sin x + b\cos x = c\sin(x + \varphi)$ with $c = \sqrt{a^2 + b^2}$.

Lemma 5.1. "Simplified" Gram[y=0] and Gram[x=0] points-Dirichlet eta functions are derived directly from Dirichlet eta function with Euler formula application and (respectively) they will intrinsically incorporate actual location [but not actual positions] of all Gram[y=0] and Gram[x=0] points.

Proof. For Gram[y=0] points, the equivalent of Eq. (4) and Eq. (6) are respectively given by Eq. (16) and Eq. (17) below.

$$\sum ReIm\{\eta(s)\} = Re\{\eta(s)\} + 0, \text{ or simply } Im\{\eta(s)\} = 0 \tag{16}$$

$$\sum_{n=1}^{\infty}(2n)^{-\sigma}\sin(t\ln(2n)) = \sum_{n=1}^{\infty}(2n-1)^{-\sigma}\sin(t\ln(2n-1))$$

$$\sum_{n=1}^{\infty}(2n)^{-\sigma}\sin(t\ln(2n)) - \sum_{n=1}^{\infty}(2n-1)^{-\sigma}\sin(t\ln(2n-1)) = 0 \tag{17}$$

For Gram[x=0] points, the equivalent of Eq. (4) and Eq. (6) are respectively given by Eq. (18) and Eq. (19) below.

$$\sum ReIm\{\eta(s)\} = 0 + Im\{\eta(s)\}, \text{ or simply } Re\{\eta(s)\} = 0 \tag{18}$$

$$\sum_{n=1}^{\infty}(2n)^{-\sigma}\cos(t\ln(2n)) = \sum_{n=1}^{\infty}(2n-1)^{-\sigma}\cos(t\ln(2n-1))$$

$$\sum_{n=1}^{\infty}(2n)^{-\sigma}\cos(t\ln(2n)) - \sum_{n=1}^{\infty}(2n-1)^{-\sigma}\cos(t\ln(2n-1)) = 0 \tag{19}$$

Eq. (17) and Eq. (19) being the "simplified" Gram[y=0] and Gram[x=0] points-Dirichlet eta functions derived directly from $\eta(s)$ will intrinsically incorporate *actual location [but not actual positions]* of (respectively) all Gram[y=0] and Gram[x=0] points. *The proof is now complete for Lemma 5.1*□.

Proposition 5.2. Gram[y=0] and Gram[x=0] points-Dirichlet Sigma-Power Laws in continuous (integral) format given as equations and inequations can both be (respectively) derived directly from "simplified" Gram[y=0] and Gram[x=0] points-Dirichlet eta functions in discrete (summation) format with Riemann integral application. [Note: Gram[y=0] and Gram[x=0] points-Dirichlet Sigma-Power Laws in continuous (integral) format here refers to relevant end-products obtained from "first key step of converting Riemann zeta function into its continuous format version".]

Proof. Antiderivatives below using (2n) parameter help obtain all subsequent equations: first two for Gram[y=0] points and second two for Gram[x=0] points.

$$\int_{1}^{\infty}(2n)^{-\sigma}\sin(t\ln(2n))dn =$$

$$\left[-\frac{(2n)^{1-\sigma}\left((\sigma-1)\sin\left(t\ln\left(2n\right)\right)+t\cos\left(t\ln\left(2n\right)\right)\right)}{2\left(t^2+(\sigma-1)^2\right)}+C\right]_1^\infty$$

$$\int_1^\infty \sin(t\ln(2n))dn = \left[\frac{(2n)(\sin\left(t\ln\left(2n\right)\right)-t\cos\left(t\ln\left(2n\right)\right))}{2(t^2+1)}+C\right]_1^\infty$$

$$\int_1^\infty (2n)^{-\sigma}\cos(t\ln(2n))dn = \left[\frac{(2n)^{1-\sigma}\left(t\sin\left(t\ln\left(2n\right)\right)-(\sigma-1)\cos\left(t\ln\left(2n\right)\right)\right)}{2\left(t^2+(\sigma-1)^2\right)}+C\right]_1^\infty$$

$$\int_1^\infty \cos(t\ln(2n))dn = \left[\frac{(2n)(t\sin\left(t\ln\left(2n\right)\right)+\cos\left(t\ln\left(2n\right)\right))}{2(t^2+1)}+C\right]_1^\infty$$

For Gram[y=0] points-Dirichlet Sigma-Power Law, the equivalent of Eq. (9) and Eq. (11) are respectively given by Eq. (20) as equation and Eq. (21) as inequation.

$$-\frac{1}{2\left(t^2+(\sigma-1)^2\right)}\cdot\left[(2n)^{1-\sigma}\left((\sigma-1)\sin\left(t\ln\left(2n\right)\right)+t\cos\left(t\ln\left(2n\right)\right)\right)-\right.$$

$$\left.(2n-1)^{1-\sigma}\left((\sigma-1)\sin\left(t\ln\left(2n-1\right)\right)+t\cos\left(t\ln\left(2n-1\right)\right)\right)\right]_1^\infty = 0 \tag{20}$$

$$\left[\frac{(2n)(\sin\left(t\ln\left(2n\right)\right)-t\cos\left(t\ln\left(2n\right)\right))}{(2n-1)(\sin\left(t\ln\left(2n-1\right)\right)-t\cos\left(t\ln\left(2n-1\right)\right))}-\frac{(2n)^{\sigma+1}}{(2n-1)^{\sigma+1}}\right]_1^\infty \neq 0 \tag{21}$$

For Gram[x=0] points-Dirichlet Sigma-Power Law, the equivalent of Eq. (9) and Eq. (11) are respectively given by Eq. (22) as equation and Eq. (23) as inequation.

$$\frac{1}{2\left(t^2+(\sigma-1)^2\right)}\cdot\left[(2n)^{1-\sigma}\left(t\sin\left(t\ln\left(2n\right)\right)-(\sigma-1)\cos\left(t\ln\left(2n\right)\right)\right)-\right.$$

$$\left.(2n-1)^{1-\sigma}\left(t\sin\left(t\ln\left(2n-1\right)\right)-(\sigma-1)\cos\left(t\ln\left(2n-1\right)\right)\right)\right]_1^\infty = 0 \tag{22}$$

$$\left[\frac{(2n)(t\sin\left(t\ln\left(2n\right)\right)+\cos\left(t\ln\left(2n\right)\right))}{(2n-1)(t\sin\left(t\ln\left(2n-1\right)\right)+\cos\left(t\ln\left(2n-1\right)\right))}-\frac{(2n)^{\sigma+1}}{(2n-1)^{\sigma+1}}\right]_1^\infty \neq 0 \tag{23}$$

Intended derivation of Gram[y=0] and Gram[x=0] points-Dirichlet Sigma-Power Laws as equations and inequations is successful. *The proof is now complete for Lemma 5.2*□.

Proposition 5.3. Exact Dimensional analysis homogeneity at $\sigma = \frac{1}{2}$ in Gram[y=0] and Gram[x=0] points-Dirichlet Sigma-Power Laws as equations and inequations are (respectively) indicated by $\sum$(all fractional exponents) = whole number '1' and '3'.

Proof. Gram[y=0] points-Dirichlet Sigma-Power Law as equation for $\sigma = \frac{1}{2}$ value is given by: $-\frac{1}{2t^2+\frac{1}{2}}\cdot\left[(2n)^{\frac{1}{2}}\left(t\cos(t\ln(2n))-\frac{1}{2}\sin(t\ln(2n))\right)-\right.$

$$\left.(2n-1)^{\frac{1}{2}}\left(t\cos(t\ln(2n-1))-\frac{1}{2}\sin(t\ln(2n-1))\right)\right]_1^\infty = 0 \tag{24}$$

"Last two terms" at n = 1:
$$\frac{1}{2t^2+\frac{1}{2}}[(2(t^2+\frac{1}{4}))^{\frac{1}{2}}\sin\left((t\ln 2)-\tan^{-1}(2t)\right)-(t^2+\frac{1}{4})^{\frac{1}{2}}\sin\left(\tan^{-1}(2t)\right)]$$

In Figure 10, the phenomenon of "monotonously decreasing height waves of varying amplitude

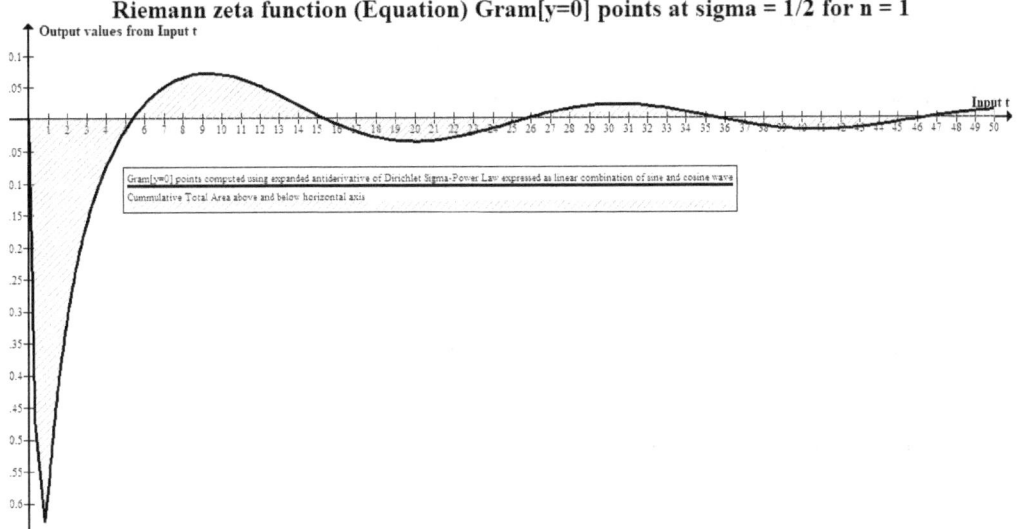

FIGURE 10. Gram[y=0] points-Dirichlet Sigma-Power Law as equation for $\sigma = \frac{1}{2}$ at n = 1.

with different n values" comply with functions that are odd occurring at $\sigma = \frac{1}{2}$.
Gram[y=0] points-Dirichlet Sigma-Power Law as inequation for $\sigma = \frac{1}{2}$ value is given by:

$$\left[\frac{(2n)(\sin(t\ln(2n)) - t\cos(t\ln(2n)))}{(2n-1)(\sin(t\ln(2n-1)) - t\cos(t\ln(2n-1)))} - \frac{(2n)^{\frac{3}{2}}}{(2n-1)^{\frac{3}{2}}}\right]_1^\infty \neq 0 \qquad (25)$$

"Last two terms" at n = 1 (as inequation): $-\dfrac{(2)\left((t^2+1)^{\frac{1}{2}}\sin((t\ln 2) - \tan^{-1}(t))\right)}{(1)\left((t^2+1)^{\frac{1}{2}}\sin((t\ln 1) - \tan^{-1}(t))\right)} + \dfrac{(2)^{\frac{3}{2}}}{(1)^{\frac{3}{2}}}$

In Figure 11, the phenomenon of "constant height waves of varying amplitude with different n

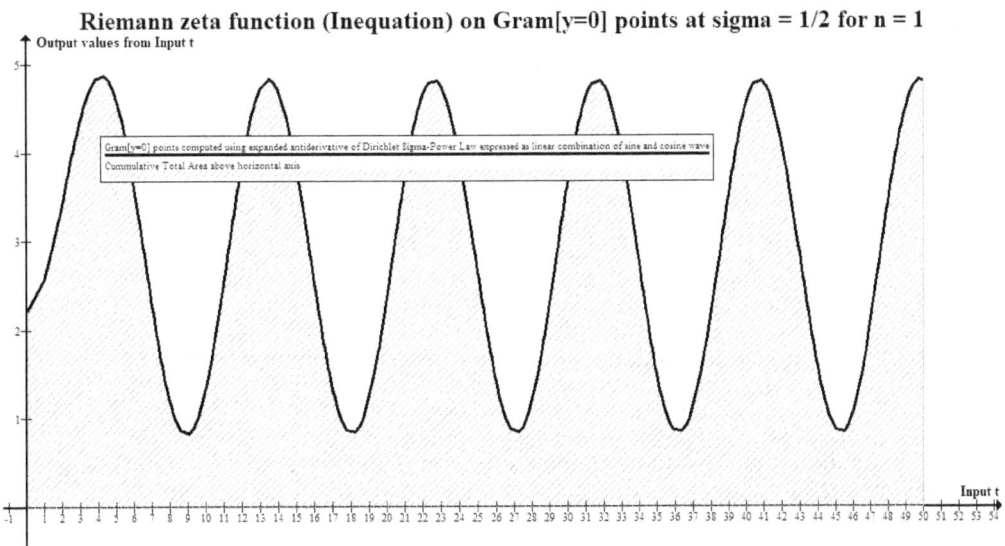

FIGURE 11. Gram[y=0] points-Dirichlet Sigma-Power Law as inequation for $\sigma = \frac{1}{2}$ at n = 1.

values" comply with functions that are even occurring at $\sigma = \frac{2}{5}$.

Gram[x=0] points-Dirichlet Sigma-Power Law as equation for $\sigma = \frac{1}{2}$ value is given by:

$$\frac{1}{2t^2+\frac{1}{2}} \cdot [(2n)^{\frac{1}{2}} \left(t\sin(t\ln(2n)) + \frac{1}{2}\cos(t\ln(2n))\right) -$$

$$(2n-1)^{\frac{1}{2}}\left(t\sin(t\ln(2n-1)) + \frac{1}{2}\cos(t\ln(2n-1))\right)]_1^\infty = 0 \qquad (26)$$

"Last two terms" at n = 1:
$$\frac{1}{2t^2+\frac{1}{2}}[-(2(t^2+\frac{1}{4}))^{\frac{1}{2}}\sin\left((t\ln 2)-\tan^{-1}(\frac{1}{2t})\right)+(t^2+\frac{1}{4})^{\frac{1}{2}}\sin\left(\tan^{-1}(\frac{1}{2t})\right)]$$

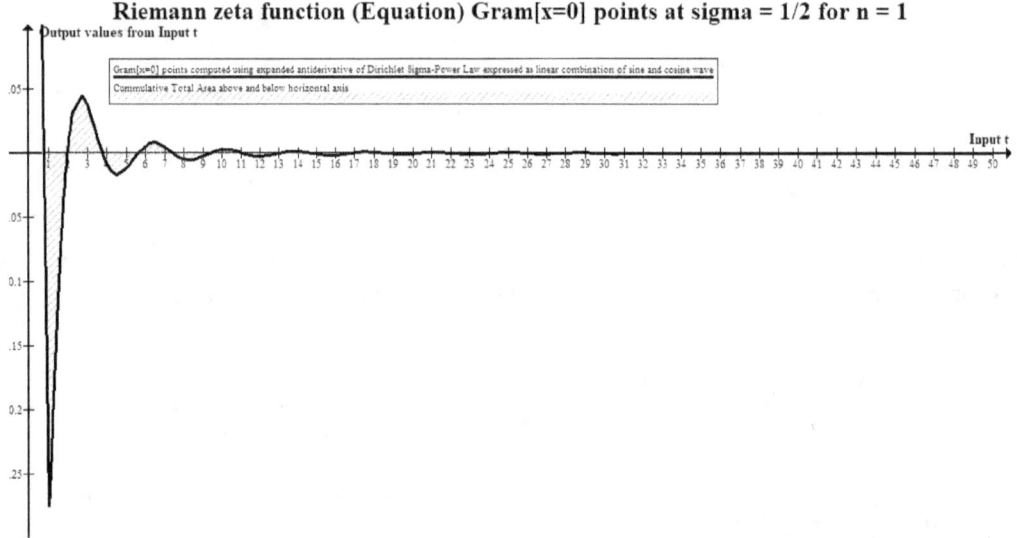

FIGURE 12. Gram[x=0] points-Dirichlet Sigma-Power Law as equation for $\sigma = \frac{1}{2}$ at n = 1.

In Figure 12, the phenomenon of "monotonously decreasing height waves of varying amplitude with different n values" comply with functions that are odd occurring at $\sigma = \frac{1}{2}$.

Gram[x=0] points-Dirichlet Sigma-Power Law as inequation for $\sigma = \frac{1}{2}$ value is given by:

$$\left[\frac{(2n)(t\sin(t\ln(2n))+\cos(t\ln(2n)))}{(2n-1)(t\sin(t\ln(2n-1))+\cos(t\ln(2n-1)))} - \frac{(2n)^{\frac{3}{2}}}{(2n-1)^{\frac{3}{2}}}\right]_1^\infty \neq 0 \qquad (27)$$

"Last two terms" at n = 1 (as inequation): $-\dfrac{(2)\left((t^2+1)^{\frac{1}{2}}\sin((t\ln 2)+\tan^{-1}(\frac{1}{t}))\right)}{(1)\left((t^2+1)^{\frac{1}{2}}\sin((t\ln 1)+\tan^{-1}(\frac{1}{t}))\right)} + \dfrac{(2)^{\frac{3}{2}}}{(1)^{\frac{3}{2}}}$

In Figure 13, the phenomenon of "monotonously increasing height waves of varying amplitude with different n values" comply with functions that are even occurring at $\sigma = \frac{2}{5}$.

$\sum$(all fractional exponents) as $2(1-\sigma)$ = whole number '1' for Eqs. (24) and (26), and $2(\sigma+1)$ = whole number '3' for Eqs. (25) and (27). These findings signify presence of complete

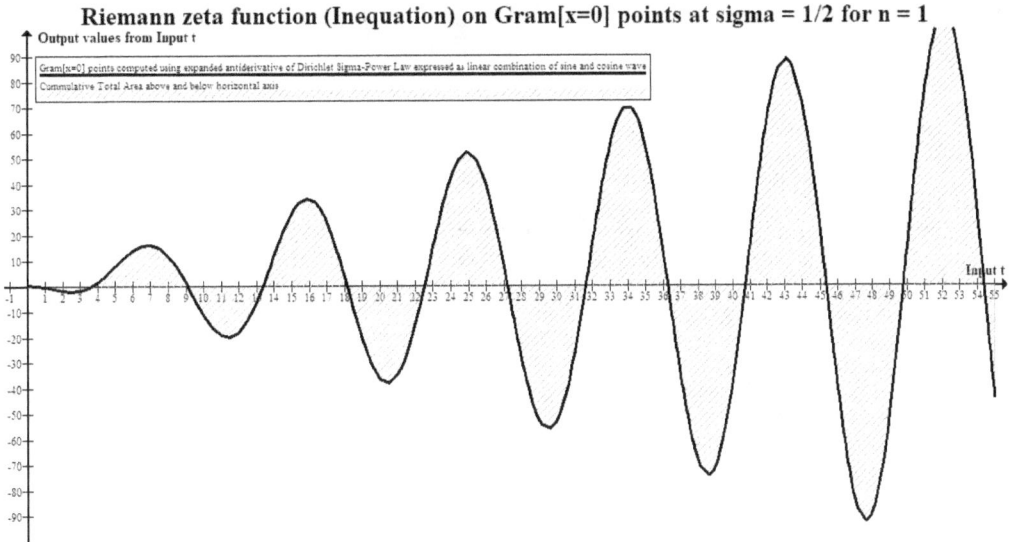

FIGURE 13. Gram[x=0] points-Dirichlet Sigma-Power Law as inequation for $\sigma = \frac{1}{2}$ at n = 1.

sets Gram[y=0] points for Eqs. (24) and (25) and Gram[x=0] points for Eqs. (26) and (27). *The proof is now complete for Proposition 5.3*□.

Corollary 5.4. Inexact Dimensional analysis homogeneity at $\sigma \neq \frac{1}{2}$ [illustrated using $\sigma = \frac{2}{5}$] in Gram[y=0] and Gram[x=0] points-Dirichlet Sigma-Power Laws as equations and inequations are (respectively) indicated by $\sum$(all fractional exponents) = fractional number '$\neq 1$' and '$\neq 3$'.

Proof. Gram[y=0] points-Dirichlet Sigma-Power Law as equation for $\sigma = \frac{2}{5}$ value is given by: $-\frac{1}{2t^2 + \frac{18}{25}} \cdot [(2n)^{\frac{3}{5}} \left(t \cos(t \ln(2n)) - \frac{3}{5} \sin(t \ln(2n)) \right) -$

$$(2n-1)^{\frac{3}{5}} \left(t \cos(t \ln(2n-1)) - \frac{3}{5} \sin(t \ln(2n-1)) \right)]_1^\infty = 0 \tag{28}$$

"Last two terms" at n = 1:
$\frac{1}{2t^2 + \frac{18}{25}} [(2^{\frac{3}{5}}(t^2 + \frac{9}{25}))^{\frac{1}{2}} \sin\left((t \ln 2) - \tan^{-1}(\frac{5t}{3}) \right) - (t^2 + \frac{9}{25})^{\frac{1}{2}} \sin\left(\tan^{-1}(\frac{5t}{3}) \right)]$

In Figure 14, the phenomenon of "monotonously decreasing height waves of varying amplitude with different n values" comply with functions that are odd occurring at $\sigma \neq \frac{1}{2}$.

Gram[y=0] points-Dirichlet Sigma-Power Law as inequation for $\sigma = \frac{2}{5}$ value is given by:

$$\left[\frac{(2n)(\sin(t \ln(2n)) - t \cos(t \ln(2n)))}{(2n-1)(\sin(t \ln(2n-1)) - t \cos(t \ln(2n-1)))} - \frac{(2n)^{\frac{7}{5}}}{(2n-1)^{\frac{7}{5}}} \right]_1^\infty \neq 0 \tag{29}$$

"Last two terms" at n = 1 (as inequation): $-\frac{(2)\left((t^2+1)^{\frac{1}{2}} \sin((t \ln 2) - \tan^{-1}(t)) \right)}{(1)\left((t^2+1)^{\frac{1}{2}} \sin((t \ln 1) - \tan^{-1}(t)) \right)} + \frac{(2)^{\frac{7}{5}}}{(1)^{\frac{7}{5}}}$

In Figure 15, the phenomenon of "constant height waves of varying amplitude with different n values" comply with functions that are even occurring at $\sigma \neq \frac{1}{2}$.

Gram[x=0] points-Dirichlet Sigma-Power Law as equation for $\sigma = \frac{2}{5}$ value is given by: $\frac{1}{2t^2 + \frac{18}{25}} \cdot$

©Professor Bernhard (Pseudonym) Riemann, viXra, Wednesday 15 January 2020

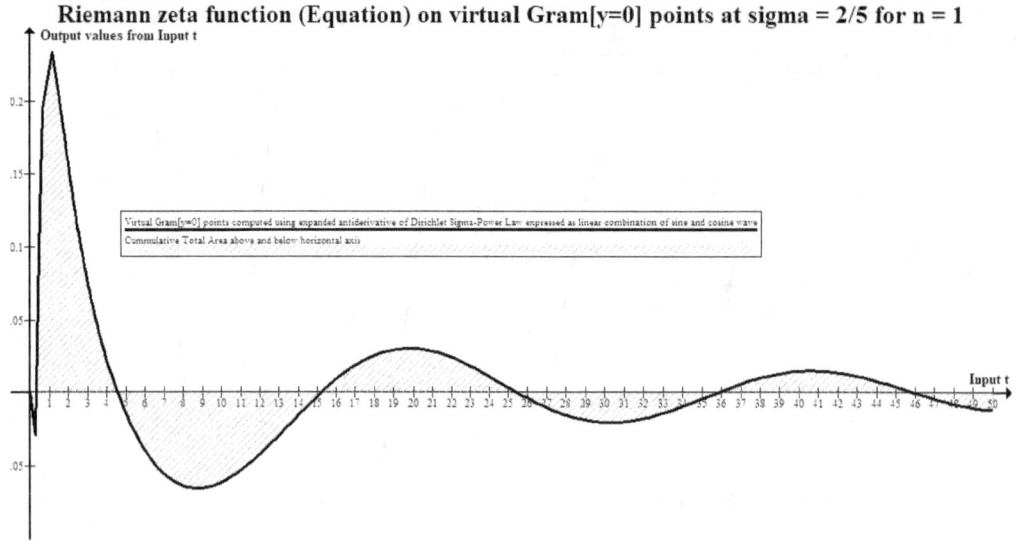

FIGURE 14. Gram[y=0] points-Dirichlet Sigma-Power Law as equation for $\sigma = \frac{2}{5}$ at n = 1.

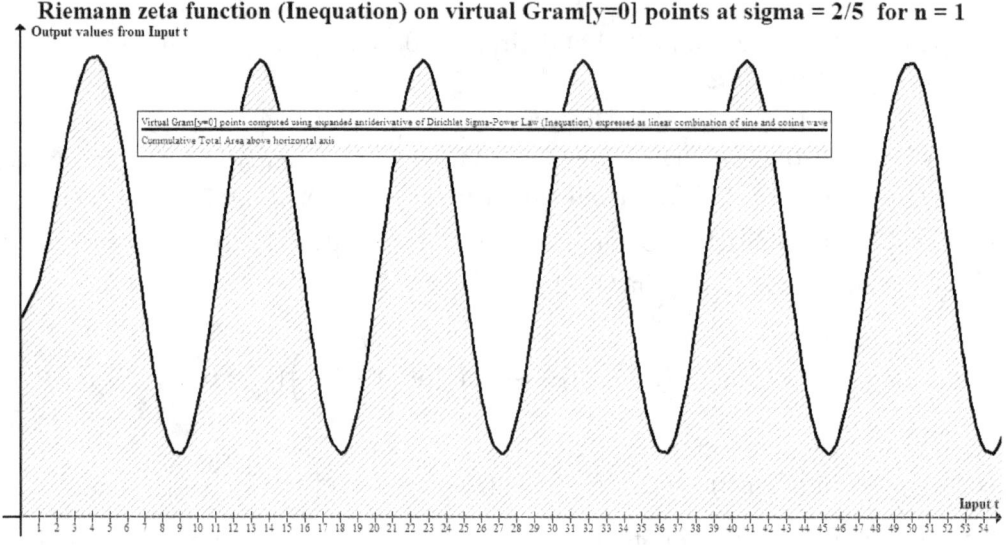

FIGURE 15. Gram[y=0] points-Dirichlet Sigma-Power Law as inequation for $\sigma = \frac{2}{5}$ at n = 1.

$$[(2n)^{\frac{3}{5}}\left(t\sin(t\ln(2n)) + \frac{3}{5}\cos(t\ln(2n))\right) -$$

$$(2n-1)^{\frac{3}{5}}\left(t\sin(t\ln(2n-1)) + \frac{3}{5}\cos(t\ln(2n-1))\right)]_1^\infty = 0 \qquad (30)$$

"Last two terms" at n = 1:
$$\frac{1}{2t^2 + \frac{18}{25}}[-(2^{\frac{3}{5}}(t^2 + \frac{9}{25}))^{\frac{1}{2}}\sin\left((t\ln 2) - \tan^{-1}(\frac{3}{5t})\right) + (t^2 + \frac{9}{25})^{\frac{1}{2}}\sin\left(\tan^{-1}(\frac{3}{5t})\right)]$$

In Figure 16, the phenomenon of "monotonously decreasing height waves of varying amplitude with different n values" comply with functions that are odd occurring at $\sigma \neq \frac{1}{2}$.

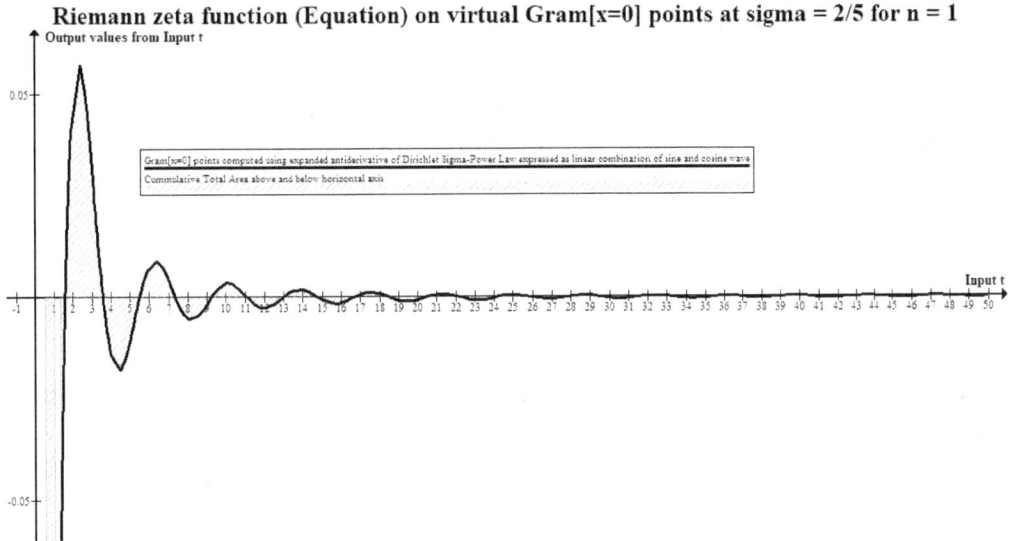

FIGURE 16. Gram[x=0] points-Dirichlet Sigma-Power Law as equation for $\sigma = \frac{2}{5}$ at n = 1.

Gram[x=0] points-Dirichlet Sigma-Power Law as inequation for $\sigma = \frac{2}{5}$ value is given by:

$$\left[\frac{(2n)(t\sin(t\ln(2n)) + \cos(t\ln(2n)))}{(2n-1)(t\sin(t\ln(2n-1)) + \cos(t\ln(2n-1)))} - \frac{(2n)^{\frac{7}{5}}}{(2n-1)^{\frac{7}{5}}}\right]_1^\infty \neq 0 \qquad (31)$$

"Last two terms" at n = 1 (as inequation): $-\dfrac{(2)\left((t^2+1)^{\frac{1}{2}}\sin((t\ln 2) + \tan^{-1}(\frac{1}{t}))\right)}{(1)\left((t^2+1)^{\frac{1}{2}}\sin((t\ln 1) + \tan^{-1}(\frac{1}{t}))\right)} + \dfrac{(2)^{\frac{7}{5}}}{(1)^{\frac{7}{5}}}$

In Figure 17, the phenomenon of "monotonously increasing height waves of varying amplitude

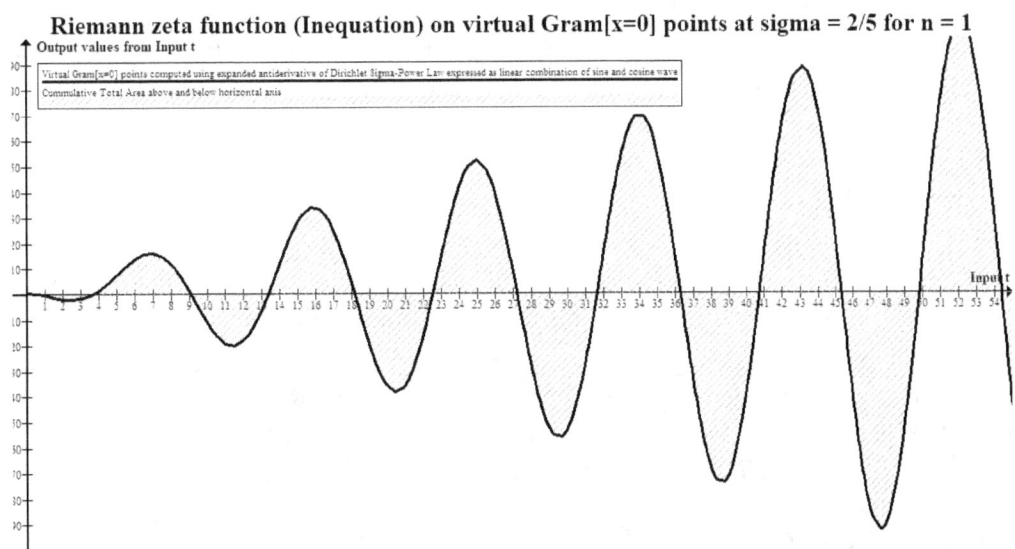

FIGURE 17. Gram[x=0] points-Dirichlet Sigma-Power Law as inequation for $\sigma = \frac{2}{5}$ at n = 1.

with different n values" comply with functions that are even occurring at $\sigma \neq \frac{1}{2}$.

$\sum$(all fractional exponents) as $2(1 - \sigma)$ = fractional number '$\neq 1$' for Eqs. (28) and (30), and $2(\sigma + 1)$ = fractional number '$\neq 3$' for Eqs. (29) and (31). These findings signify presence of complete sets virtual Gram[y=0] points for Eqs. (28) and (29) and virtual Gram[x=0] points for Eqs. (30) and (31). *The proof is now complete for Corollary 5.4*$\square$.

6. Prime and Composite numbers

Prime & Composite numbers are Incompletely Predictable entities dependently linked together in a sequential, cummulative & eternal manner since the relationship Number '1' + Prime numbers + Composite numbers = Natural numbers always hold for all Natural numbers.

6.1 Dimensional analysis on Cardinality and "Dimensions" for Prime numbers

We use the word "Dimensions" to denote well-defined Incompletely Predictable entities obtained from Information-Complexity conservation. Relevant "Dimensions" *dependently* represent Number '1', **P** and **C**. Then *by default* any (sub)sets of **P** and **C** in well-defined equations can also be represented by their corresponding "Dimensions".

Remark 6.1. We can apply Dimensional analysis to "Dimensions" from Information-Complexity conservation and cardinality of relevant sets in certain well-defined equations.

Let **X** denote **E**, **O**, **N** [which are classified as Completely Predictable numbers], **P** and **C** [which are classified as Incompletely Predictable numbers]. For x = 1, 2, 3, 4, 5,..., ∞; consider all **X** $\leq$ x. Then this "all **X** $\leq$ x" is definition for **X**-$\pi(x)$ [denoting "**X** counting function"] resulting in following two types of equations coined as (I) 'Exact' equation **N**-$\pi(x)$ = **E**-$\pi(x)$ + **O**-$\pi(x)$ with "non-varying" relationships **E**-$\pi(x)$ = **O**-$\pi(x)$ for all x = **E** and **E**-$\pi(x)$ = **O**-$\pi(x)$ - 1 for all x = **O**, and (II) 'Inexact' equation **N**-$\pi(x)$ = 1 + **P**-$\pi(x)$ + **C**-$\pi(x)$ with "varying" relationships **P**-$\pi(x)$ > **C**-$\pi(x)$ for all x $\leq$ 8; **P**-$\pi(x)$ = **C**-$\pi(x)$ for x = 9, 11, and 13; and **P**-$\pi(x)$ < **C**-$\pi(x)$ for x = 10, 12, and all x $\geq$ 14.

Let "Dimensions" and different (sub)sets of **E**, **O**, **N**, **P** and **C** be 'base quantities'. Then exponent '1' of "Dimensions" and cardinality of these (sub)sets in well-defined equations are corresponding 'units of measurement'. Performing DA on "Dimensions" for **PC** pairing is depicted later on. Performing DA on cardinality is depicted next.

For Set **N** = Set **E** + Set **O**, then $|\mathbf{N}| = |\mathbf{E}| + |\mathbf{O}| \implies \aleph_0 = \aleph_0 + \aleph_0$ thus conforming with DA homogeneity.

For Set **N** = Set **P** + Set **C** + Number '1', then Set **N** - Number '1' = Set **P** + Set **C** and $|\mathbf{N}$ - Number '1'$| = |\mathbf{P}| + |\mathbf{C}| \implies \aleph_0 = \aleph_0 + \aleph_0$ thus conforming with DA homogeneity.

For Set **N** - Set **even P** - Number '1'= Set **odd P** + Set **even C** + Set **odd C**, then $|\mathbf{N}$ - **even P** - Number '1'$|$ = $|$odd **P**$|$ + $|$even **C**$|$ + $|$odd **C**$| \implies \aleph_0 = \aleph_0 + \aleph_0 + \aleph_0$ thus conforming with DA homogeneity. Symbolically represented by all available **O** prime gap = 1 and **E** prime gaps = 2, 4, 6, 8, 10,...; **O** composite gap = 1 and **E** composite gap = 2; and **O** natural gap = 1; then $|\mathbf{Gap\ 1\ N} - \mathbf{Gap\ 1\ P} - $ Number '1'$| = |\mathbf{Gap\ 2\ P}| + |\mathbf{Gap\ 4\ P}| + |\mathbf{Gap\ 6\ P}| + |\mathbf{Gap\ 8\ P}| + |\mathbf{Gap\ 10\ P}| + ... + |\mathbf{Gap\ 1\ C}| + |\mathbf{Gap\ 2\ C}| \implies \aleph_0 = \aleph_0 + \aleph_0 + \aleph_0 + \aleph_0 + \aleph_0 + ... \aleph_0 + \aleph_0$ thus conforming with DA homogeneity. It is known that $|\mathbf{Gap\ 1\ P}| = |$Number '2'$| = 1$ and $|\mathbf{Gap\ 1\ N}| = |\mathbf{Gap\ 1\ C}| = |\mathbf{Gap\ 2\ C}| = \aleph_0$. Then solving Polignac's and Twin prime conjectures translate to successfully proving |**Gap 2 P**| = |**Gap 4 P**| = |**Gap 6 P**| = |**Gap 8 P**| = |**Gap 10 P**| = ... = $\aleph_0$ with |**E prime gaps**| = $\aleph_0$.

Outline of proof for Polignac's and Twin prime conjectures. Requires simultaneously satisfying two mutually inclusive conditions: I. *With rigid manifestation of DA homogeneity*, quantitive[7] fulfillment by considering i ∈ **E** for each Subset **odd** P_i generated by **E** prime gap = i from Set **E prime gaps** occurs only if solitary cardinality value is present in equation Set **odd P** = $\sum_{i=2}^{\infty}$ Subset **odd** P_i with |**odd P**| = |**odd** P_i| = |**E prime gaps**| = $\aleph_0$, and II. *With rigid manifestation of DA non-homogeneity*, quantitive[7] fulfillment by considering i ∈ **E** for each Subset **odd** P_i generated by **E** prime gap = i from Set **E prime gaps** does not occur if more than one cardinality values are present in equation Set **odd P** > $\sum_{i=2}^{\infty}$ Subset **odd** P_i with |**E prime gaps**| = $\aleph_0$ having incorrect |Subset(s) **odd P**| = N (finite value) &/or Set **odd P** > $\sum_{i=2}^{N}$ Subset **odd** P_i with |**odd** P_i| = $\aleph_0$ having incorrect |**E prime gaps**| = N (finite value).

Footnote 7: Qualitative fulfillment of |**odd P**| = |**odd** P_i| = |all **E prime gaps**| = $\aleph_0$ equates to Plus-Minus Gap 2 Composite Number Alternating Law being precisely obeyed by all **E** prime gaps apart from first **E** prime gap precisely obeying Plus Gap 2 Composite Number Continuous Law. Derived using Information-Complexity conservation, these Laws symbolize "end-result" proof on Polignac's and Twin prime conjectures. *Law of Continuity* is a heuristic principle *whatever succeed for the finite, also succeed for the infinite*. Then these Laws which inherently manifest 'Gap 2 Composite Number' on finite and infinite time scale should in principle "succeed for the finite, also succeed for the infinite".

Polignac's and Twin prime conjectures mathematical foot-prints. Six identifiable steps to prove these conjectures: *Step 1* Let N = 2x - ΣPC$_x$-Gap. Define Dimension (2x - N) based on Information-Complexity conservation to validly represent **P** & **C**. Considering x ∈ **N**, obtain Dimensions (2x - 2), (2x - 4), (2x - 5), (2x - 7), (2x - 8), (2x - 9), ..., (2x - ∞) with specific groupings to constitute all elements of Set **P** [culminating in obtaining all prime gaps (= **E** prime gaps + Solitary **O** prime gap) with |**all prime gaps**| = $\aleph_0$]. Note Dimension (2x - 2) represents x = 1 (Number '1') which is neither **P** nor **C**. *Step 2* Considering i ∈ **E**, confirm perpetual recurrences of individual **E** prime gap = i (associated with its unique odd P_i) occur only when depicted as specific groupings of Dimension (2x - N)^1 now endowed with exponent '1' for all ranges of x. *Step 3* Perform DA on exponent '1' in these Dimensions. *Step 4* Perform DA on equation Set **odd P** = $\sum_{i=2}^{\infty}$ Subset **odd** P_i to obtain |**odd P**| = |**odd** P_i| = $\aleph_0$ whereby Subset **odd** P_i is derived from its associated unique **E** prime gap = i with |**E prime gaps**| = $\aleph_0$. *Step 5* Confirm 'Prime number' variable and 'Prime gap' variable complex algorithm "containing" all **P** with knowing their overall actual location [but not actual positions][8]. *Step 6* Derive Plus-Minus Gap 2 Composite Number Alternating Law and Plus Gap 2 Composite Number Continuous Law using Dimension (2x - N)^1 and Information-Complexity conservation.

Footnote 8: This phrase implies all **P** (and **C**) are Incompletely Predictable numbers. Actual positions will require using complex algorithm Sieve of Eratosthenes to *dependently* calculate positions of all preceding **P** (and **C**) in neighborhood – see Lemma 8.4 [P_{n+1} = 2 + $\sum_{i=1}^{n} G_{Pi}$

with '2' denoting $\mathbf{P}_1$] & Lemma 8.5 [$\mathbf{C}_{n+1} = 4 + \sum_{i=1}^{n} \mathbf{G}_{Ci}$ with '4' denoting $\mathbf{C}_1$].

'Complex Elementary Fundamental Laws'-based solutions of Plus-Minus Gap 2 Composite Number Alternating Law and Plus Gap 2 Composite Number Continuous Law are obtained by undertaking the non-negotiable mathematical steps outlined above. These Laws are literally Completely Predictable meta-properties ('overall' *complex properties*) arising from "interactions" between $\mathbf{P}$ and $\mathbf{C}$ producing relevant patterns of Gap 2 Composite Number perpetual appearances [albeit with Incompletely Predictable timing]. We logically deduce explicit mathematical explanation for these meta-properties requires "complex" mathematical arguments. Attempts to give explicit mathematical explanation with "simple" mathematical arguments would intuitively mean Incompletely Predictable numbers $\mathbf{P}$ and $\mathbf{C}$ be (incorrectly and impossibly) treated as Completely Predictable numbers.

6.2 Brief overview of Polignac's and Twin prime conjectures

Occurring over 2000 years ago (c. 300 BC), ancient Euclid's proof on infinitude of $\mathbf{P}$ in totality [viz. $|\mathbf{P}| = \aleph_0$ for Set $\mathbf{P}$] predominantly by *reductio ad absurdum* (proof by contradiction) is earliest known but not the only proof for this simple problem in Number theory. Since then dozens of proofs have been devised such as three chronologically listed: Goldbach's Proof using Fermat numbers (written in a letter to Swiss mathematician Leonhard Euler, July 1730), Furstenberg's Topological Proof in 1955[5], and Filip Saidak's Proof in 2006[6]. The strangest candidate is likely to be Furstenberg's Topological Proof.

In 2013, Yitang Zhang proved a landmark result showing some unknown even number 'N' < 70 million such that there are infinitely many pairs of $\mathbf{P}$ that differ by 'N'[7]. By optimizing Zhang's bound, subsequent Polymath Project collaborative efforts using a new refinement of GPY sieve in 2013 lowered 'N' to 246; and assuming Elliott-Halberstam conjecture and its generalized form have further lower 'N' to 12 and 6, respectively. Then 'N' has intuitively more than one valid values such that there are infinitely many pairs of $\mathbf{P}$ that differ by each of those 'N' values [thus proving existence of more than one Subset **odd** $\mathbf{P}_i$ with |**odd** $\mathbf{P}_i$| = $\aleph_0$]. We can only theoretically lower 'N' to 2 (in regards to $\mathbf{P}$ with 'small gaps') but there are still an infinite number of $\mathbf{E}$ prime gaps (in regards to $\mathbf{P}$ with 'large gaps') that require "the proof that each will generate its unique set of infinite $\mathbf{P}$".

Remark 6.2. Existence of maximal and non-maximal prime gaps supply crucial indirect evidence to intuitively support but does not prove "Each even prime gap will generate an infinite magnitude of odd prime numbers on its own accord".

Comments relevant to Remark 6.2 are given in the next section below.

7. Supportive role of maximal and non-maximal prime gaps

We analyze data of all $\mathbf{P}$ obtained when extrapolated out over a wide range of x ⩾ 2 integer values. As sequence of $\mathbf{P}$ carries on, $\mathbf{P}$ with ever larger prime gaps will appear. For given range of x integer values, prime gap = n_2 is a 'maximal prime gap' if prime gap = n_1 < prime gap = n_2 for all $n_1 < n_2$. In other words, the largest such prime gaps in this range are called maximal prime gaps. The term 'first occurrence prime gaps' refers to first occurrences of maximal prime gaps whereby maximal prime gaps are prime gaps of "at least of this length". We use maximal prime gaps to denote 'first occurrence prime gaps'. CIS non-maximal prime gaps (endorsed with nickname 'slow jumpers') will always lag behind CIS maximal prime gaps for onset appearances

TABLE 1. First 17 prime gaps depicted in the format utilizing maximal prime gaps [depicted with asterisk symbol (*)] and non-maximal prime gaps [depicted without asterisk symbol].

Prime gap	Following prime number	Prime gap	Following prime number
1*	2	18*	523
2*	3	20*	887
4*	7	22*	1129
6*	23	24	1669
8*	89	26	2477
10	139	28	2971
12	199	30	4297
14*	113	32	5591
16	1831	..	....

in **P** sequence. These are shown for first 17 prime gaps in Table 1. Apart from **O** prime gap = 1 representing solitary even **P** '2', remaining **P** in Table 1 consist of representative single odd **P** for each **E** prime gap. These odd **P** individually make one-off appearance in **P** sequence in a *perpetual albeit Incompletely Predictable manner*. Initial seven of [majority] "missing" odd **P** are 5, 11, 13, 17, 19, 29, 31,... belonging to Subset **P** with 'residual' prime gaps are potential source of odd **P** in relation to proposal that each **E** prime gap from Set **E prime gaps** will generate its specific Subset **odd P**. Set **all P** from all prime gaps = Subset **P** from maximal prime gaps + Subset **P** from non-maximal prime gaps + Subset **P** from 'residual' prime gaps. Subset **P** from 'residual' prime gaps with representation from all **E** prime gaps must include all correctly selected "missing" odd **P**. These observations support but does not prove proposition that each **E** prime gap will generate its own Subset **odd P** with $|\text{odd } \mathbf{P}| = \aleph_0$.

For $i \in \mathbf{N}$; primordial $P_i\#$ is analog of usual factorial for **P** = 2, 3, 5, 7, 11, 13,.... Then $P_1\#$ = 2, $P_2\#$ = 2 X 3 = 6, $P_3\#$ = 2 X 3 X 5 = 30, $P_4\#$ = 2 X 3 X 5 X 7 = 210, $P_5\#$ = 2 X 3 X 5 X 7 X 11 = 2310, $P_6\#$ = 2 X 3 X 5 X 7 X 11 X 13 = 30030, etc. English mathematician John Horton Conway coined the term 'jumping champion' in 1993. An integer n is a 'jumping champion' if n is the most frequently occurring difference (prime gap) between consecutive **P**<x for some x integer values. Example: for any x with 7<x<131, n = 2 (indicating twin **P**) is the 'jumping champion'. It has been conjectured that (i) the only 'jumping champions' are 1, 4 and primorials 2, 6, 30, 210, 2310, 30030,... and (ii) 'jumping champions' tend to infinity. Their required proofs will likely need proof of k-tuple conjecture. **P** from 'jumping champion' prime gaps have their onset appearances in **P** sequence in a *perpetual albeit Incompletely Predictable manner* [as another example to that outlined in previous paragraph].

8. Information-Complexity conservation

A formula, as equation or algorithm, is simply a Black Box generating necessary Output (with qualitative structural 'Complexity') when supplied with given Input (with quantitative data 'Information'). This 'Information' and 'Complexity' are what is referred to in the term 'Information-Complexity conservation'. **P** and **C** numbers are traditionally "analyzed separately". The key definition behind Dimension $(2x - N)^1$ is used to abstractly represent *dependent* **P** and **C** numbers (and Number '1') in a combined manner whereby N = 2x - ΣPC_x-Gap. This will lead to required mathematical arguments based on Information-Complexity conservation and patterns in Gap 2 Composite Number to obtain Plus-Minus Gap 2 Composite Number Alternating Law and Plus Gap 2 Composite Number Continuous Law which will [respectively] solve Polignac's and Twin prime conjectures.

Obvious analogy: Just as there is conservation or preservation of "net area value" happening at appropriate times for continuous format Riemann zeta function (*aka* Dirichlet Sigma-Power Law); similar conservation or preservation of "net number value" will happen at appropriate times for natural numbers on one hand and prime numbers, composite numbers and Number '1' on the other hand when Information-Complexity conservation is enforced. This concept will be equally applicable to natural numbers, prime numbers, composite numbers and Number '1' when depicted using Dimension (2x - N)[1].

N (CIS): 1, 2, 3,..., $+\infty$. Let x be from Set **X** such that x $\in$ **N**. Consider x for upper boundary of interest in Set **X** whereby **X** is chosen from **N**, **E**, **O**, **P** or **C**.

Lemma 8.1. Natural counting function **N**-$\pi(x)$, defined as $|\mathbf{N} \leqslant \mathrm{x}|$, is Completely Predictable by independently using simple algorithm to be equal to x.

Proof Formula to generate **N** with 100% certainty is $\mathbf{N}_i = i$ whereby $\mathbf{N}_i$ is the ith **N** and i = 1, 2, 3,..., ∞. For a given $\mathbf{N}_i$, its ith position is simply i. Natural gap ($\mathbf{G}_{Ni}$) = $\mathbf{N}_{i+1}$ - $\mathbf{N}_i$, with $\mathbf{G}_{Ni}$ always = 1. There are x **N** $\leqslant$ x. Thus **N**-$\pi(x) = |\mathbf{N} \leqslant \mathrm{x}| = $ x. *The proof is now complete for Lemma 8.1*□.

Lemma 8.2. Even counting function **E**-$\pi(x)$, defined as $|\mathbf{E} \leqslant \mathrm{x}|$, is Completely Predictable by independently using simple algorithm to be equal to floor(x/2).

Proof. Formula to generate **E** with 100% certainty is $\mathbf{E}_i = $ iX2 whereby $\mathbf{E}_i$ is the ith **E** and i = 1, 2, 3,..., ∞ abiding to mathematical label "All **N** always ending with a digit 0, 2, 4, 6 or 8". For a given $\mathbf{E}_i$, its ith position is calculated as i = $\mathbf{E}_i$/2. Even gap ($\mathbf{G}_{Ei}$) = $\mathbf{E}_{i+1}$ - $\mathbf{E}_i$, with $\mathbf{G}_{Ei}$ always = 2. There are $\lfloor \frac{x}{2} \rfloor$ **E** $\leqslant$ x. Thus **E**-$\pi(x) = |\mathbf{E} \leqslant \mathrm{x}| = $ floor(x/2). *The proof is now complete for Lemma 8.2*□.

Lemma 8.3. Odd counting function **O**-$\pi(x)$, defined as $|\mathbf{O} \leqslant \mathrm{x}|$, is Completely Predictable by independently using simple algorithm to be equal to ceiling(x/2).

Proof. Formula to generate **O** with 100% certainty is $\mathbf{O}_i = $ (iX2) - 1 whereby $\mathbf{O}_i$ is the ith odd number and i = 1, 2, 3,..., ∞ abiding to mathematical label "All **N** always ending with a digit 1, 3, 5, 7, or 9". For a given $\mathbf{O}_i$ number, its ith position is calculated as i = ($\mathbf{O}_i$ + 1)/2. Odd gap ($\mathbf{G}_{Oi}$) = $\mathbf{O}_{i+1}$ - $\mathbf{O}_i$, with $\mathbf{G}_{Oi}$ always = 2. There are $\lceil \frac{x}{2} \rceil$ **O** $\leqslant$ x. Thus **O**-$\pi(x) = |\mathbf{O} \leqslant \mathrm{x}| = $ ceiling(x/2). *The proof is now complete for Lemma 8.3*□.

Lemma 8.4. Prime counting function **P**-$\pi(x)$, defined as $|\mathbf{P} \leqslant \mathrm{x}|$, is Incompletely Predictable with Set **P** dependently obtained using complex algorithm Sieve of Eratosthenes.

Proof. Algorithm to generate $\mathbf{P}_i$ whereby $\mathbf{P}_1$ (= 2), $\mathbf{P}_2$ (= 3), $\mathbf{P}_3$ (= 5), $\mathbf{P}_4$ (= 7),..., ∞ with 100% certainty is based on Sieve of Eratosthenes abiding to mathematical label "All **N** apart from 1 that are evenly divisible by itself and by 1". Although we can check primality of a given **O** by trial division, we can never determine its position without knowing positions of preceding **P**. Prime gap ($\mathbf{G}_{Pi}$) = $\mathbf{P}_{i+1}$ - $\mathbf{P}_i$, with $\mathbf{G}_{Pi}$ constituted by all **E** except 1st $\mathbf{G}_{P1}$ = 3 - 2 = 1. **P**-$\pi(x) = |\mathbf{P} \leqslant \mathrm{x}|$. This is Incompletely Predictable and is calculated via mentioned algorithm. Using definition of prime gap, every **P** [represented here with aid of 'n' notation instead of usual 'i' notation] is written as $\mathbf{P}_{n+1} = 2 + \sum_{i=1}^{n} \mathbf{G}_{Pi}$ with '2' denoting $\mathbf{P}_1$. Here i & n = 1, 2, 3, 4, 5, ..., ∞. *The proof is now complete for Lemma 8.4*□.

Lemma 8.5. Composite counting function **C**-$\pi(x)$, defined as $|\mathbf{C} \leqslant \mathrm{x}|$, is Incompletely Predictable with Set **C** derived as Set **N** - Set **P** [dependently obtained using complex algorithm Sieve of Eratosthenes] - Number '1'.

Proof. Composite numbers abide to mathematical label "All **N** apart from 1 that are evenly divisible by numbers other than itself and 1". Algorithm to generate C_i whereby C_1 (= 4), C_2 (= 6), C_3 (= 8), C_4 (= 9),..., ∞ with 100% certainty is based [indirectly] on Sieve of Eratosthenes via selecting non-prime **N** to be **C**. We define Composite gap G_{Ci} as C_{i+1} - C_i with G_{Ci} constituted by 1 & 2. **C**-$\pi(x)$ = |**C** ⩽ x|. This is Incompletely Predictable and need to be calculated indirectly via the mentioned algorithm. Using definition of composite gap, every **C** [represented here with aid of 'n' notation instead usual 'i' notation] is written as $C_{n+1} = 4 + \sum_{i=1}^{n} G_{Ci}$ with '4' denoting C_1. Here i & n = 1, 2, 3, 4, 5, ..., ∞. *The proof is now complete for Lemma 8.5*□.

Denote **X** to be **N**, **E**, **O**, **P** or **C**. **X**-$\pi(x)$ = |**X** ⩽ x| with x ∈ **N**. We define and compute entity 'Grand-Total Gaps for **X** at x' (Grand-Total ΣX_x-Gaps).

Proposition 8.6. For any given x ⩾ 1 values in Set **N**, designated Complexity is represented by ΣN_x-Gaps = x - N with N = 1 being maximal.

Proof. Set **N** (for x = 1 to 12): 1, 2, 3, 4, 5, 6, 7, 8, 9, 10, 11, 12. **N**-$\pi(x)$ = 12. There are x - 1 = 11 **N**-Gaps each of '1' magnitude: 1, 1, 1, 1, 1, 1, 1, 1, 1, 1, 1. ΣN_x-Gaps = 11 X 1 = 11. This equates to "x - 1" – regarded as Complexity for **N**. *The proof is now complete for Proposition 8.6*□.

Proposition 8.7. For any given x ⩾ 1 values in constituent Set **E** and Set **O**, designated Complexity is represented by ΣEO_x-Gaps = 2x - N with N = 4 being maximal.

Proof. Set **E** and Set **O** (for x = 1 to 12): 2, 4, 6, 8, 10, 12 and 1, 3, 5, 7, 9, 11. **E**-$\pi(x)$ = 6 and **O**-$\pi(x)$ = 6. There are $\lfloor \frac{x}{2} \rfloor$ - 1 = 5 **E**-Gaps each of '2' magnitude: 2, 2, 2, 2, 2. ΣE_x-Gaps = 5 X 2 = 10, and $\lceil \frac{x}{2} \rceil$ - 1 = 5 **O**-Gaps each of '2' magnitude: 2, 2, 2, 2, 2. ΣO_x-Gaps = 5 X 2 = 10. Grand-Total ΣEO_x-Gaps = 10 + 10 = 20. Depicted by Table 3 and Figure 19 in Appendix D, 2x - N = "2x - 4" [perpetual constant appearances of "N = 4 being maximal"] is Complexity for **E** and **O**. *The proof is now complete for Proposition 8.7*□.

Proposition 8.8. For selected x ⩾ 2 values in constituent Set **P** and Set **C**, designated Complexity is cyclically represented by ΣPC_x-Gaps = 2x - N with N = 7 being minimal.

Proof. Set **P** and Set **C** (for x = 2 to 12): 2, 3, 5, 7, 11 and 4, 6, 8, 9, 10, 12. **P**-$\pi(x)$ = 5 and **C**-$\pi(x)$ = 6. There are four **P**-Gaps of 1, 2, 2, 4 magnitude and five **C**-Gaps of 2, 2, 1, 1, 2 magnitude. ΣP_x-Gaps = 1 + 2 + 2 + 4 = 9. ΣC_x-Gaps = 2 + 2 + 1 + 1 + 2 = 8. Grand-Total ΣPC_x-Gaps = 9 + 8 = 17. Depicted by Table 2 and Figure 18, 2x - N = "2x - 7" [perpetual intermittent and cyclical appearances of "N = 7 being minimal"] is Complexity for **P** and **C**. *The proof is now complete for Proposition 8.8*□.

Designated Complexity is (i) x - N with N = 1 (maximal) for Completely Predictable **N**, (ii) 2x - N with N = 7 (minimal) for Incompletely Predictable **P** & **C**, and (iii) 2x - N with N = 4 (maximal) for Completely Predictable **E** & **O**. Interpretations: **N** has minimal Complexity, **E** & **O** have intermediate Complexity, and **P** & **C** have maximal [varying] Complexity. Defacto baseline "2x - 4" Grand-Total Gaps [minus 4 value] in **E**-**O** pairing > Defacto baseline "2x - ⩾7" Grand-Total Gaps [minus ⩾7 values] in **P**-**C** pairing.

Let both x & N ∈ **N**. We tabulate in Table 2 and graph in Figure 18 [Incompletely Predictable] **P**-**C** mathematical landscape for a relatively larger x = 2 to 64 here (and ditto for [Completely Predictable] **E**-**O** mathematical landscape for relatively larger x = 1 to 64 in Appendix D). The term "mathematical landscape" denotes specific mathematical patterns in tabulated and graphed data. "Dimension" contextually denotes Dimension 2x - N whereby (i) allocated [infinite] N values result in Dimensions 2x - 7, 2x - 8, 2x - 9, ..., 2x - ∞ for **P-C** finite scale mathematical landscape and (ii) allocated [finite] N values for **E-O** finite scale mathematical landscape result in Dimension 2x - 4. For **P-C** pairing, initial one-off Dimensions 2x - 2, 2x - 4 and 2x - 5 (in consecutive order) are exceptions [with Dimension 2x - 2 validly representing Number '1' which is neither **P** nor **C**]. For **E-O** pairing, initial one-off Dimension 2x - 2 is an exception. **P-C** mathematical landscape consisting of Dimensions will intrinsically incorporate **P** and **C** in an integrated manner and there are infinite times whereby relevant Dimensions deviate away from 'baseline' Dimension 2x - 7 simply because **P** [and, by default, **C**] in totality are rigorously proven to be infinite in magnitude. In contrast, there is a complete lack of deviation away from 'baseline' Dimension 2x - 4 apart from one-off deviation caused by the initial Dimension 2x - 2 in Appendix D.

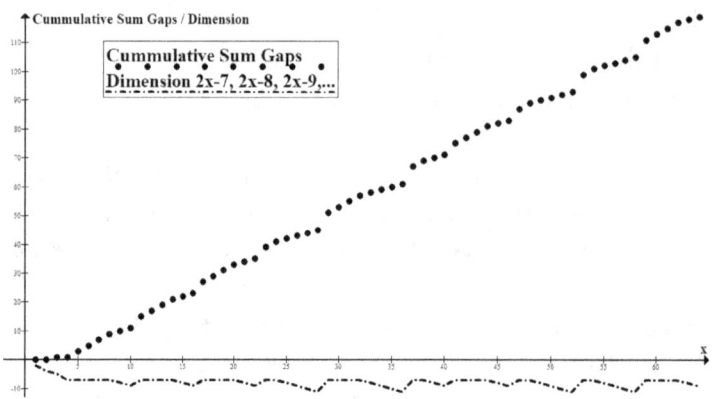

FIGURE 18. Prime-Composite finite scale mathematical (graphed) landscape using data obtained for x = 2 to 64. Bottom graph symbolically represent "Dimensions" using ever larger negative integers.

In Figure 18, Dimensions 2x - 7, 2x - 8, 2x - 9, ..., 2x - ∞ are symbolically represented by -7, -8, -9, ..., ∞ with 2x - 7 displayed as 'baseline' Dimension whereby Dimension trend (Cumulative Sum Gaps) must repeatedly reset itself onto this 'baseline' Dimension on a perpetual basis. Dimensions symbolically represented by ever larger negative integers will correspond to **P** associated with ever larger prime gaps and this phenomenon will generally happen at ever larger x values (with complete presence of Chaos and Fractals being manifested in our graph). At ever larger x values, **P**-$\pi(x)$ will overall become larger but with a *decelerating* trend whereas **C**-$\pi(x)$ will overall become larger but with an *accelerating* trend. This support ever larger prime gaps appearing at ever larger x values.

Definitive derivation of data in Table 2 is illustrated by two examples for position x = 31 & 32. For i & x ∈ **N**; ΣPC_x-Gap = ΣPC_{x-1}-Gap + Gap value at P_{i-1} or Gap value at C_{i-1} whereby (i) P_i or C_i at position x is determined by whether relevant x value belongs to a **P** or **C**, and (ii) both ΣPC_1-Gap and ΣPC_2-Gap = 0. Example, for position x = 31: 31 is **P** (**P11**). Desired Gap value at **P10** = 2. Thus ΣPC_{31}-Gap (55) = ΣPC_{30}-Gap (53) + Gap value at **P10** (2). Example, for position x = 32: 32 is **C** (**C20**). Desired Gap value at **C19** = 2. Thus ΣPC_{32}-Gap (57) = ΣPC_{31}-Gap (55) + Gap value at **C20** (2). Note: in our Dimension (2x - N) system, N = 2x - ΣPC_x-Gap. 'Overall magnitude of **C** will always be greater than that of **P**' will hold true from x = 14 onwards. For instance, position x = 61 corresponds to **P** 61 which is 18th **P**, whereas [one lower] position x = 60 corresponding to **C** 60 is [much higher] 42nd **C**.

TABLE 2. Prime-Composite finite scale mathematical (tabulated) landscape using data obtained for x = 2 to 64. The Number '1' is neither prime nor composite.
Legend: C = composite, P = prime, Dim = Dimension, Y = Dimension 2x - 7 (for visual clarity), N/A = Not Applicable.

x	P_i or C_i Gaps	ΣPC_x-Gaps	Dim	x	P_i or C_i Gaps	ΣPC_x-Gaps	Dim
1	N/A	0	2x-2	33	C21, 1	58	2x-8
2	P1, 1	0	2x-4	34	C22, 1	59	2x-9
3	P2, 2	1	2x-5	35	C23, 1	60	2x-10
4	C1, 2	1	Y	36	C24, 2	61	2x-11
5	P3, 2	3	Y	37	P12, 4	67	Y
6	C2, 2	5	Y	38	C25, 1	69	Y
7	P4, 4	7	Y	39	C26, 1	70	2x-8
8	C3, 1	9	Y	40	C27, 1	71	2x-9
9	C4, 1	10	2x-8	41	P13, 2	75	Y
10	C5, 2	11	2x-9	42	C28, 2	77	Y
11	P5, 2	15	Y	43	P14, 4	79	Y
12	C6, 2	17	Y	44	C29, 1	81	Y
13	P6, 4	19	Y	45	C30, 1	82	2x-8
14	C7, 1	21	Y	46	C31, 2	83	2x-9
15	C8, 1	22	2x-8	47	P15, 6	87	Y
16	C9, 1	23	2x-9	48	C32, 1	89	Y
17	P7, 2	27	Y	49	C33, 1	90	2x-8
18	C10, 2	29	Y	50	C34, 1	91	2x-9
19	P8, 4	31	Y	51	C35, 1	92	2x-10
20	C11, 1	33	Y	52	C36, 1	93	2x-11
21	C12, 1	34	2x-8	53	P16, 6	99	Y
22	C13, 2	35	2x-9	54	C37, 1	101	Y
23	P9, 6	39	Y	55	C38, 1	102	2x-8
24	C14, 1	41	Y	56	C39, 1	103	2x-9
25	C15, 1	42	2x-8	57	C40, 1	104	2x-10
26	C16, 1	43	2x-9	58	C41, 1	105	2x-11
27	C17, 1	44	2x-10	59	P17, 2	111	Y
28	C18, 2	45	2x-11	60	C42, 2	113	Y
29	P10, 2	51	Y	61	P18, 6	115	Y
30	C19, 2	53	Y	62	C43, 1	117	Y
31	P11, 6	55	Y	63	C44, 1	118	2x-8
32	C20, 1	57	Y	64	C45, 1	119	2x-9

9. Polignac's and Twin prime conjectures

Previous section alludes to **P-C** finite scale mathematical landscape. This section alludes to **P-C** infinite scale mathematical landscape. Let 'Y' symbolizes (baseline) Dimension 2x - 7. Let prime gap at $\mathbf{P}_i = \mathbf{P}_{i+1} - \mathbf{P}_i$ with $\mathbf{P}_i$ & $\mathbf{P}_{i+1}$ respectively symbolizes consecutive "first" & "second" **P** in any $\mathbf{P}_i$-$\mathbf{P}_{i+1}$ pairings. We denote (i) Dimensions YY grouping [depicted by 2x - 7 initially appearing twice in (iii)] to represent signal for appearances of **P** pairings other than twin **P** such as cousin **P**, sexy **P**, etc; (ii) Dimension YYYY grouping to represent signal for appearances of **P** pairings as twin **P**; and (iii) Dimension (2x - $\geqslant$7)-Progressive-Grouping allocated to 2x - 7, 2x - 7, 2x - 8, 2x - 9, 2x - 10, 2x - 11,..., 2x - ∞ as elements of *precise* and *proportionate* CFS Dimensions representation of an individual $\mathbf{P}_i$ with its associated prime gap namely, Dimensions 2x - 7 & 2x - 7 pairing = twin **P** (with both its prime gap & CFS cardinality = 2); 2x - 7, 2x - 7, 2x - 8 & 2x - 9 pairing = cousin **P** (with both its prime gap & CFS cardinality = 4); 2x - 7, 2x - 7, 2x - 8, 2x - 9, 2x - 10 & 2x - 11 pairing = sexy **P** (with both its prime gap & CFS cardinality = 6); and so on. The higher order [traditionally defined as closest possible] prime groupings of three **P** as **P** triplets, of four **P** numbers as prime quadruplets, of five **P** numbers as prime quintuplets, etc consist of serendipitous groupings abiding to mathematical rule: With exception of three 'outlier' **P** 3, 5, & 7; groupings of any three **P** as **P**, **P**+2, **P**+4 combination (viz. manifesting two consecutive twin **P**) is a mathematical impossibility. The 'anomaly' one of every three consecutive **O** is a multiple of three, and hence this number cannot be **P**, explains this impossibility. Then closest possible **P** grouping [viz. for prime triplet] must be either **P**, **P**+2, **P**+6 or **P**, **P**+4, **P**+6 format.

P groupings not respecting traditional closest-possible-prime groupings are also the norm occurring infinitely often, indicating continual presence of prime gaps $\geqslant$ 6. As **P** become sparser at larger range, perpetual presence of (i) prime gaps $\geqslant$ 6 [proposed by us to arbitrarily represent 'large gaps'] and (ii) prime gaps 2 & 4 [proposed by us to arbitrarily represent 'small gaps'] with progressive greater magnitude will cummulatively occur for each prime gap but always in a decelerating manner. With permanent requirement at larger range of intermittently resetting to baseline Dimension 2x - 7 occurring [either two or] four times in a row, nature seems to dictate, at the very least, perpetual twin **P** or one other non-twin **P** occurrences is inevitable.

We dissect Dimension YYYY unique signal for twin **P** appearances: Initial two CFS Dimensions YY components of YYYY represent "first" **P** component of twin **P** pairing. Last two Dimensions YY components of YYYY signifying appearance of "second" **P** component of twin **P** pairing is also the initial first-two-element component of full CFS Dimensions representation for "first" **P** component of following non-twin **P** pairing. Twin **P** are uniquely represented by repeating *single* type Dimension 2x - 7. In all other 'higher order' **P** pairings (with prime gaps $\geqslant$ 4), they require *multiple* types Dimension representation. There is qualitative aspect association of *single* type Dimension representation for twin **P** resulting in "less colorful" Plus Gap 2 Composite Number *Continuous Law* as opposed to *multiple* types Dimension representation for all other 'higher order' **P** pairings resulting in "more colorful" Plus-Minus Gap 2 Composite Number *Alternating Law*. 'Gap 2 Composite Number' occurrences in both Laws on finite scale are (directly) observed in Figure 18 & Table 2 for x = 2 to 64, and on infinite scale are (indirectly) deduced using logical arguments for all x values.

We endow all "Dimensions" with exponent of '1' for perusal in on-going mathematical arguments. $\mathbf{P}_1 = 2$ is represented by CFS as Dimension $(2x - 4)^1$ (with both prime gap & CFS cardinality = 1); $\mathbf{P}_2 = 3$ is represented by CFS as Dimensions $(2x - 5)^1$ & $(2x - 7)^1$ (with both prime gap & CFS cardinality = 2); $P_3 = 5$ is represented by CFS Dimension $(2x - 7)^1$ & $(2x - 7)^1$ (with both prime gap & CFS cardinality = 2), etc.

Proposition 9.1. Let Case 1 be Completely Predictable **E** & **O** pairing and Case 2 be Incompletely Predictable **P** & **C** pairing. Furthermore, let Case 1 and Case 2 be independent of each other. Then for any given x value, there exist grand total number of Dimensions [Complexity] such that it exactly equal to either two combined subtotal number of Dimensions [Complexity] to precisely represent **E** & **O** in Case 1, or combined subtotal number of Dimensions [Complexity] to precisely represent **P** & **C** & Number '1' in Case 2.

Proof. **N** is directly constituted from either combined **E** & **O** in Case 1 or combined **P** & **C** & Number '1' in Case 2 – Number '1' is neither **P** nor **C**. Correctly designated infinitely many CFS of Dimensions used to represent combined **E** & **O** in Case 1 and combined **P** & **C** & Number '1' in Case 2 must also directly and proportionately be representative of relevant **N** arising from combined subtotal of **E** & **O** in Case 1 and from combined subtotal of **P** & **C** & Number '1' in Case 2. *The proof is now complete for Proposition 9.1*□.

Proposition 9.2. Let Case 1 be Completely Predictable **E** & **O** pairing and Case 2 be Incompletely Predictable **P** & **C** pairing. Furthermore, let Case 1 and Case 2 be independent of each other. Part I: For any given x value apart from x = 1 value in Case 1 and x = 1, 2, and 3 values in Case 2; Dimension $(2x - N)^1$ [Complexity] representations of all Completely Predictable **E** & **O** in Case 1 and all Incompletely Predictable **P** & **C** & Number '1' in Case 2 are given by N = 4 in Case 1 and by N $\geqslant$ 7 in Case 2. Part II: Odd **P** obeys Plus-Minus Composite Gap 2 Number Alternating Law for prime gaps $\geqslant$ 4 and Plus Composite Gap 2 Number Continuous Law for prime gap = 2.

Proof. Apart from first Dimension $(2x - 2)^1$ representation in **E** & **O** pairing in Case 1 and first three Dimension $(2x - 2)^1$, Dimension $(2x - 4)^1$ and Dimension $(2x - 5)^1$ representations in **P** & **C** pairing in Case 2; possible N value in Dimension $(2x - N)^1$ representation are shown to be (constantly) maximal 4 for Case 1 and (variably) minimal 7 for Case 2. For Case 2, we again note Dimension $(2x - 2)^1$ to (validly) represent Number '1' which is neither **P** nor **C**. These nominated Dimensions simply represent possible (constant) baseline "2x - 4" Grand-Total Gaps as per Proposition 8.7 for Case 1 & (variable) baseline "2x - 7" Grand-Total Gaps as per Proposition 8.8 for Case 2. Note that all CFS of Dimensions that can be used to precisely represent combined **E** & **O** in Case 1 will persistently consist of same [solitary] Dimension $(2x - 4)^1$ after first Dimension $(2x - 2)^1$. Perpetual repeated deviation of N values away from N = 7 (minimum) in Case 2 is simply representing infinite magnitude of **P** & **C**. *The proof is now complete for Part I of Proposition 9.2*□.

Derived Dimensions will comply with Incompletely Predictable property as explained using **P** '61'. At Position x = 61 equating to $P_{18} = 61$, it is represented by CFS Dimensions $(2x - 7)^1$, $(2x - 7)^1$, $(2x - 8)^1$, $(2x - 9)^1$, $(2x - 10)^1$ & $(2x - 11)^1$ (with both prime gap & CFS cardinality = 6). This representation indicates an "unknown but correct" **P** with prime gap = 6 when we intentionally conceal full information '61' = 31st **O** = 18th **P** with prime gap = 6. But to arrive at this representation requires calculations of all preceding CFS Dimensions thus manifesting hallmark Incompletely Predictable property of CFS Dimensions.

Overall sum total of individual CFS Dimensions required to represent every **P** is infinite in magnitude as |**all P**| = $\aleph_0$. Standalone Dimensions YY groupings [representing signals for "higher order" non-twin **P** appearances] &/or as front Dimensions YY (sub)groupings [which by itself is fully representative of twin **P** as Dimensions YYYY appearances] need to recur on an indefinite basis. Then twin **P** and "higher order" cousin **P**, sexy **P**, etc should aesthetically all be infinite in magnitude because (respectively) they regularly and universally arise as part of Dimension YYYY and Dimension YY appearances. An isolated **P** is defined as a **P** such that neither **P** - 2 nor **P** + 2 is **P**. In other words, isolated **P** is not part of a twin **P** pair. Example 23 is an isolated **P** since 21 and 25 are both **C**. Then repeated inevitable presence of Dimension YY grouping is nothing more than indicating repeated occurrences of isolated **P**. This constitutes another view on Dimension YY.

CIS of Gap 1 Composite Numbers are fully associated with non-twin **P** as they eternally occur in between any two consecutive non-twin **P**. CIS of Gap 2 Composite Numbers are (i) fully associated with twin **P** as they are eternally present in between any twin **P** pair, and (ii) partially associated with non-twin **P** as they are eternally present alternatingly or intermittently in between any two consecutive non-twin **P**. Then (i) Gap 1 Composite Numbers do not have valid representation by **E** prime gap = 2, and (ii) Gap 2 Composite Numbers have valid representations by all **E** prime gaps = ["consistently" only for] 2, ["inconsistently" for each of] 4, 6, 8, 10,.... This is an alternative view on **P** from perspective of CFS composite gaps [instead of CIS prime gaps] with intrinsic patterns having *alternating presence* and *absence* of Gap 2 Composite Numbers associated with every CFS Dimensions representations of **P** with prime gaps ⩾ 4, viz. 'Plus-Minus Gap 2 Composite Number Alternating Law'. CFS Dimensions representations of Twin **P** are always associated with Gap 2 Composite Numbers, viz. 'Plus Gap 2 Composite Number Continuous Law'.

Examples for both Laws: A twin **P** (prime gap = 2) in its unique CFS Dimensions format always has Gap 2 Composite Numbers in a [constant] pattern. A cousin **P** (prime gap = 4) in its unique CFS Dimensions format always has two Gap 1 Composite Numbers & then one Gap 2 Composite Number [combined] pattern *alternating* with three consecutive Gap 1 Composite Numbers [non-combined] pattern. From this simple observation alone, we deduce we can generate an infinite magnitude of **C** from each composite gaps 1 & 2. Gap 2 Composite Numbers *alternating* pattern behavior in cousin **P** will not hold true unless twin **P** & all other non-cousin **P** are infinite in magnitude and integratedly supplying essential "driving mechanism" to eternally sustain this Gap 2 Composite Numbers *alternating* pattern behavior in cousin **P**. Thus we establish twin **P** and cousin **P** in their CFS Dimensions formats are CIS intertwined together when depicted using **C** with composite gaps = 1 & 2 with each supplying their own peculiar (infinite) share of associated Gap 2 Composite Numbers [thus contributing to overall pool of Gap 2 Composite Numbers].

An inevitable statement in relation to "Gap 2 Composite Numbers pool contribution" based on above reasoning: At the bare minimum, *either* twin **P** *or* at least one of non-twin **P** must be infinite in magnitude. An inevitable impression: All generated subsets of **P** from 'small gaps' [of 2 & 4] and 'large gaps' [of ⩾ 6] alike should each be CIS thus allowing true uniformity in **P** distribution. Again we see in Table 2 depicting **P-C** data for x = 2 to 64 that, for instance, **P** with prime gap = 6 must also persistently have this 'last-place' Gap 2 Composite Numbers intermittently appearing in certain rhythmic *alternating* patterns, thus complying with Plus-Minus Gap 2 Composite Number Alternating Law. This CFS Dimensions representation for **P** with prime gaps = 6 will again generate their infinite share of associated Gap 2 Composite Numbers to contribute to this pool. The presence of this last-place Gap 2 Composite Numbers in various alternating pattern in appearances & non-appearances must *self-generatingly* be similarly extended in a mathematically consistent fashion *ad infinitum* to all other remaining infinite number of prime gaps [which were not discussed in details above]. *The proof is now complete for Part II of Proposition 9.2*□.

10. Rigorous Proofs now named as Polignac's and Twin prime hypotheses

The proofs on lemmas and propositions from previous section supply all necessary evidences to fully support Theorem Polignac-Twin prime I to IV below thus depicting proofs for Polignac's and Twin prime conjectures in a rigorous manner. Gap 1 Composite Numbers do not have valid representation by **E** prime gap = 2, and Gap 2 Composite Numbers have valid representations by all **E** prime gaps = ["consistently" only for] 2, ["inconsistently" for each of] 4, 6, 8, 10,.... Plus-Minus Gap 2 Composite Number Alternating Law confirms that Gap 2 Composite Numbers present in each **P** with prime gaps ≥ 4 situation must appear as some sort of "rhythmic patterns of alternating presence and absence" for Gap 2 Composite Numbers. Twin **P** with prime gap = 2 obeying Plus Gap 2 Composite Number Continuous Law can be understood as special situation of "(non-)rhythmic patterns with continual presence" for relevant Gap 2 Composite Numbers.

In 1849 when French mathematician Alphonse de Polignac (1826 - 1863) was admitted to Polytechnique, he created Polignac's conjecture which relates complete set of odd **P** to all **E** prime gaps. Made earlier by de Polignac in 1846, Twin prime conjecture which relates twin prime numbers to prime gap = 2, is nothing more than a subset of Polignac's conjecture.

Theorem Polignac-Twin prime I. Incompletely Predictable prime numbers $\mathbf{P}_n = 2, 3, 5, 7, 11, ..., \infty$ or composite numbers $\mathbf{C}_n = 4, 6, 8, 9, 10, ..., \infty$ are CIS with overall actual location [but not actual positions] of all prime or composite numbers accurately represented by complex algorithm involving prime gaps G_{Pi} viz. $\mathbf{P}_{n+1} = 2 + \sum_{i=1}^{n} G_{Pi}$ or involving composite gaps G_{Ci} viz. $\mathbf{C}_{n+1} = 4 + \sum_{i=1}^{n} G_{Ci}$ whereby prime & composite numbers are symbolically represented here with aid of 'n' notation instead of usual 'i' notation; and i & n = 1, 2, 3, 4, 5, ..., ∞. Number '2' in first algorithm represents $\mathbf{P}_1$, the very first (and only even) **P**. Number '4' in second algorithm represent $\mathbf{C}_1$, the very first (and even) **C**.

Proof. We treat above algorithms as unique mathematical objects looking for key intrinsic properties and behaviors. Each **P** or **C** is assigned a unique prime or composite gap. Absolute number of **P** or **C** and (thus) prime or composite gaps are infinite in magnitude. As original formulae containing all **P** or **C** by themselves (viz. without supplying prime or composite gaps as "input information" to generate **P** or **C** as "output complexity"), these algorithms intrinsically incorporate overall actual location [but not actual positions] of all **P** or **C**. *The proof is now complete for Theorem Polignac-Twin prime I*□.

Theorem Polignac-Twin prime II. Set of prime gaps $G_{Pi} = 2, 4, 6, 8, 10, ..., \infty$ is infinite in magnitude whereby these prime gaps accurately and completely represented by Dimensions $(2x - 7)^1, (2x - 8)^1, (2x - 9)^1, ..., (2x - \infty)^1$ must satisfy Information-Complexity conservation in a consistent manner.

Proof. Part I of Proposition 9.2 proved all **P** are represented by Dimension $(2x - N)^1$ with $N \geq 7$ for any given x value (except for x = 2 & 3 values). Note that although x = 1 is neither **P** nor **C**, it is validly represented by Dimension $(2x - 2)^1$. If each **P** is endowed with a specific prime gap value, then each such prime gap must [via logical mathematical deduction] be represented by Dimension $(2x - N)^1$. We advocate this nominated method of prime gap representation using Dimensions be [purportedly] the only way to achieve Information-Complexity conservation. The preceding mathematical statements are correct as there is a unique prime gap value associated with each **P**. Proposition 10.1 below based on principles from Set theory provides further supporting materials that prime gaps are infinite in magnitude. *The proof is now complete for Theorem Polignac-Twin prime II*□.

Theorem Polignac-Twin prime III. To maintain Dimensional analysis (DA) homogeneity, those Dimensions $(2x - N)^1$ from Theorem Polignac-Twin prime II must contain eternal repetitions of well-ordered sets constituted by Dimensions $(2x - 7)^1$, $(2x - 8)^1$, $(2x - 9)^1$, $(2x - 10)^1$, $(2x - 11)^1$, ..., $(2x - \infty)^1$.

Proof. This Theorem is stated in greater details as "To maintain DA homogeneity, those aforementioned [endowed with exponent 1] Dimensions $(2x - N)^1$ from Theorem Polignac-Twin prime II must repeat themselves indefinitely in following specific combinations – (i) Dimension $(2x - 7)^1$ only appearing as twin [two-times-in-a-row] and quadruplet [four-times-in-a-row] sequences, and (ii) Dimensions $(2x - 8)^1$, $(2x - 9)^1$, $(2x - 10)^1$, $(2x - 11)^1$,..., $(2x - \infty)^1$ appearing as progressive groupings of **E** 2, 4, 6, 8, 10,..., ∞." To accommodate the only even **P** '2', exceptions to this DA homogeneity compliance will expectedly occur right at beginning of **P** sequence – (i) one-off appearance of Dimensions $(2x - 2)^1$, $(2x - 4)^1$ and $(2x - 5)^1$ and (ii) one-off appearance of Dimension $(2x - 7)^1$ as a quintuplet [five-times-in-a-row] sequence which is equivalent to (eternal) non-appearance of Dimension $(2x - 6)^1$ at x = 4. [We again note Dimension $(2x - 2)^1$ validly represent Number '1' which is neither **P** nor **C**.] These sequentially arranged sets are CFS whereby from x = 11 onwards, each set always commence initially as 'baseline' Dimension $(2x - 7)^1$ at x = **O** values and always end with its last Dimension at x = **E** values. Each set also have varying cardinality with values derived from all **E**; and correctly combined sets always intrinsically generate two infinite sets of **P** and, by default, **C** in an integrated manner. Our Theorem Polignac-Twin prime III simply represent a mathematical summary derived from Section 8 & 9 of all expressed characteristics of Dimension $(2x - N)^1$ when used to represent **P** with intrinsic display of DA homogeneity. See Proposition 10.2 for more details on DA aspect. *The proof is now complete for Theorem Polignac-Twin prime III*□.

Theorem Polignac-Twin prime IV. Aspect 1. The "quantitive" aspect to existence of both prime gaps and their associated prime numbers as sets of infinite magnitude will be shown to be correct by utilizing principles from Set theory. Aspect 2. The "qualitative" aspect to existence of both prime gaps and their associated prime numbers as sets of infinite magnitude will be shown to be correct by 'Plus-Minus Gap 2 Composite Number Alternating Law' and 'Plus Gap 2 Composite Number Continuous Law'.

Proof. Required concepts from Set theory involve cardinality of a set with its 'well-ordering principle' application. Supporting materials for these concepts based on 'pigeonhole principle' in relation to Aspect 1 are outlined in Proposition 10.1 below. 'Plus-Minus Gap 2 Composite Number Alternating Law' is applicable to all **E** prime gaps [apart from first **E** prime gap = 2 for twin primes]. The prime gap = 2 situation will obey 'Plus Gap 2 Composite Number Continuous Law'. These Laws are essentially Laws of Continuity inferring underlying intrinsic driving mechanisms that enables infinity magnitude association for both prime gaps & prime numbers to co-exist. By the same token, these Laws have the important implication that they must be applicable to those relevant prime gaps on an perpetual time scale. Supporting materials in relation to Aspect 2 are found in Proposition 9.2 above. *The proof is now complete for Theorem Polignac-Twin prime IV*□.

Note two mutually inclusive conditions: Condition 1. Presence of all Dimensions that repeat themselves on an indefinite basis and with exponent of '1' will give rise to complete sets of **P** & **C** ["DA-wise one & only one mathematical possibility argument" associated with inevitable *de novo* DA homogeneity], and Condition 2. Presence of any Dimension(s) that do not repeat itself (themselves) on an indefinite basis or with exponent other than '1' will give rise to incomplete set of **P** & **C** or incorrect set of non-**P** & non-**C** ["DA-wise mathematical impossibility argument" associated with inevitable *de novo* DA non-homogeneity]. When met, these two conditions fully support the point that CFS Dimensions representations of **P** & **C** [with respective prime & composite gaps] are totally accurate. Condition 1 reflect proof from Theorem Polignac-Twin prime III as all **P** & **C** are associated with DA homogeneity when their Dimensions are endowed with exponent of '1'. Condition 2 invoke corollary on inevitable appearance of incomplete **P** or **C** or non-**P** or non-**C** [associated with DA non-homogeneity] being tightly incorporated into this mathematical framework. See Propositions 10.1 & 10.2, and Corollary 10.3 for supporting materials on DA homogeneity & non-homogeneity.

We analyze **P** (& **C**) in terms of (i) measurements based on cardinality of CIS and (ii) pigeonhole principle which states that if n items are put into m containers, with n>m, then at least one container must contain more than one item. We note that ordinality of all infinite **P** (& **C**) is "fixed" implying that each one of the infinite well-ordered Dimension sets conforming to CFS type as constituted by Dimensions $(2x - 7)^1$, $(2x - 8)^1$, $(2x - 9)^1$, $(2x - 10)^1$, $(2x - 11)^1$, ..., $(2x - \infty)^1$ on respective gaps for **P** (& **C**) must also be "fixed".

Proposition 10.1. "Even number prime gaps are infinite in magnitude with each even number prime gap generating odd prime numbers which are again infinite in magnitude" is supported by principles from Set theory and two Laws based on Gap 2 Composite Number.

Proof. We validly exclude even **P** '2' here. Let (i) cardinality $T = \aleph_0$ for Set **all odd P** derived from **E** prime gaps 2, 4, 6,..., ∞, (ii) cardinality $T_2 = \aleph_0$ for Subset **odd P** derived from **E** prime gap 2, cardinality $T_4 = \aleph_0$ for Subset **odd P** derived from **E** prime gap 4, cardinality $T_6 = \aleph_0$ for Subset **odd P** derived from **E** prime gap 6, etc. Paradoxically, (as sets) $T = T_2 + T_4 + T_6 +... + T_\infty$ equation is valid despite (their cardinality) $T = T_2 = T_4 = T_6 =... = T_\infty$ [with well-ordering principle "stating that every non-empty set of positive integers contains a least element" fulfilled by each (sub)set]; and **E** prime gaps are 'infinite in magnitude' can justifiably be perceived instead as 'arbitrarily large in magnitude' since cumulative sum total of **E** prime gaps is relatively much slower to attain the 'infinite in magnitude' status when compared to cumulative sum total of **P** which rapidly attain this status. But if Subset **odd P** derived from one or more **E** prime gap(s) are finite in magnitude, this will breach the $\aleph_0$ cardinality 'uniformity' resulting in (i) DA non-homogeneity and (ii) inequality (as sets) $T > T_2 + T_4 + T_6 +... + T_\infty$. In language of pigeonhole principle "stating that if n items are put into m containers with n > m, then at least one container must contain more than one item", residual **odd P** (still CIS in magnitude) not accounted for by CFS-type **E** prime gap(s) will have to be [incorrectly] contained in one (or more) of composite gap(s). These arguments using cardinality constitute proof that **E** prime gaps and odd **P** generated from each **E** prime gap, are all CIS. *The proof [on "quantitative" aspect] is now complete for Proposition 10.1*□.

Complete set of **P** is represented by Dimensions $(2x - N)^1$. Table 2 & Figure 18 on **PC** finite scale mathematical landscape depict perpetual repeating features used in "qualitative" statements supporting (i) Plus-Minus Gap 2 Composite Number Alternating Law (stated as **C** with composite gaps = 2 present in each of **P** with prime gaps ≥ 4 situation must be observed to appear as some sort of rhythmic patterns of alternating presence and absence of this type of **C**), and (ii) Plus Gap 2 Composite Number Continuous Law (stated as **C** with composite gaps = 2 continual appearances in each of (twin) **P** with prime gap = 2 situation). Plus-Minus Gap 2 Composite Number Alternating Law has built-in intrinsic mechanism to automatically generate all prime gaps ≥ 4 in a mathematically consistent *ad infinitum* manner. Plus Gap 2

Composite Number Continuous Law has built-in intrinsic mechanism to automatically generate prime gap = 2 appearances in a mathematically consistent *ad infinitum* manner. *The proof [on "qualitative" aspect] is now complete for Proposition 10.1*$\square$.

Proposition 10.2. The presence of Dimensional analysis homogeneity always result in correct and complete set of prime (and composite) numbers.

Proof. DA homogeneity is completely dependent on all Dimensions being consistently endowed with exponent '1'. As all **P** (& **C**) are "fixed", we deduce from Figure 18 & Table 2 that there is one (& only one) way to represent Information-Complexity conservation using our defined Dimensions. Thus, there is one (& only one) way to depict all **P** (& **C**) using these Dimensions in a self-consistent manner and this is achieved with the one (& only one) DA homogeneity possibility. *The proof is now complete for Proposition 10.2*$\square$.

Corollary 10.3. The presence of Dimensional analysis non-homogeneity always result in incorrect and/or incomplete set of prime (and composite) numbers.

Proof. For optimal clarity, we endow all Dimensions with exponent '1' depicted as $(2x - 7)^1$, $(2x - 8)^1$, $(2x - 9)^1$, $(2x - 10)^1$, $(2x - 11)^1$,..., $(2x - \infty)^1$. Proposition 5.2 equates DA homogeneity with correct & complete set of **P** (& **C**). There are "more than one" DA possibilities when, for instance, a particular [first] term from $(2x - 7)^0$, $(2x - 8)^1$, $(2x - 9)^1$,..., $(2x - \infty)^1$ "terminates" prematurely and does not perpetually repeat [with loss of continuity]. There are intuitively two 'broad' DA possibilities here; namely, (one) DA homogeneity possibility and (one) DA non-homogeneity possibility – Dimension $(2x - 7)^0$ [$= 1$] with its exponent arbitrarily set as '0' against-all-trend in this case. Thus Dimension $(2x - 7)^1$ that stop recurring at some point in **P** (or **C**) sequence may cause well-ordered CFS sets from progressive groupings of [**E**] 2, 4, 6, 8, 10,..., ∞ for Dimensions $(2x - 8)^1$, $(2x - 9)^1$, $(2x - 10)^1$, $(2x - 11)^1$,..., $(2x - \infty)^1$ to stop existing (and ultimately for sequential **P** (or **C**) to stop appearing) at that point with ensuing outcome that **P** (or **C**) may overall be incorrectly finite or incomplete in magnitude. Finally also manifesting DA non-homogeneity, a Dimension endowed with fractional exponent values other than '1' such as '$\frac{2}{5}$' or '$\frac{3}{5}$' will result in non-**P** (or non-**C**) [fractional] numbers. *The proof is now complete for Corollary 10.3*$\square$.

Each [fixed] finite scale mathematical landscape "page" as part of [fixed] infinite scale mathematical landscape "pages" for **P** & **C** display Chaos [sensitivity to initial conditions viz. positions of subsequent **P** & **C** are "sensitive" to positions of initial **P** & **C**] and Fractals [manifesting fractal dimensions with self-similarity viz. those aforementioned Dimensions for **P** & **C** are always present, albeit in non-identical manner, for all ranges of x $\geq$ 2]. Advocated in another manner, Chaos and Fractals phenomena of those Dimensions for **P** & **C** are always present signifying accurate composition of **P** & **C** in different [predetermined] finite scale mathematical landscape "(snapshot) pages" for **P** & **C** that are self-similar but never identical – and there are an infinite number of these finite scale mathematical landscape "(snapshot) pages". The crucial mathematical step in representing all **P** (& **C**) and prime (& composite) gaps with "Dimensions" based on Information-Complexity conservation allows us to obtain the two Laws based on Gap 2 Composite Numbers and perform DA on these entities. The 'strong' principle argument is DA homogeneity equates to complete set of **P** (& **C**) whereas DA non-homogeneity does not equate to complete set of **P** (& **C**). We also advocate for a 'weak' principle argument supporting DA homogeneity for **P** (& **C**) in that nature should not "favor" any particular Dimension(s) to terminate and therefore DA non-homogeneity cannot exist for **P** (& **C**). Abiding to our advocated convention that 'conjecture' be termed 'hypothesis' once proven; we now label these conjectures as Polignac's and Twin prime hypotheses.

11. Conclusions

We mathematically envisage two *mutually exclusive* groups of entities: [totally] Unpredictable entities and [totally] Predictable entities. The first group can arise as [totally] random physical processes in nature e.g. radioactive decay is a stochastic (random) process occurring at level of single atoms. It is impossible to predict when a particular atom will decay regardless of how long the atom has existed. For a collection of atoms, expected decay rate is characterized in terms of their measured decay constants or half-lives. The second group is constituted by two subgroups: Completely Predictable entities e.g. Even-Odd number pairing in Table 3 [with abbreviation 'Y' = Dimension 2x - 4] and Incompletely Predictable entities e.g. Prime-Composite number pairing in Table 2 [with abbreviation 'Y' = Dimension 2x - 7]. **Note the [only] two common situations in both pairings from Table 2 and Table 3 whereby we identically use Dimension 2x - 2 to represent Number '1' and Dimension 2x - 4 to represent Number '2'.**

Intuitively, every single mathematical argument from complete set of mathematical arguments required to fully solve a given Incompletely Predictable Problem (containing *dependent* types of Incompletely Predictable entities) must be correct obeying Mathematics for Completely Predictable Problems. Then Mathematics for Incompletely Predictable Problems is literally the mathematical framework for describing complex properties present in these entities. For Even-Odd number pairing in Appendix D, one can [redundantly] introduce Mathematics for Completely Predictable Problems as the mathematical framework describing this Completely Predictable Problem containing *independent* types of Completely Predictable entities endowed with simple properties.

CIS of [Completely Predictable] natural numbers 1, 2, 3, 4, 5, 6, 7,... having CIS of [Completely Predictable] natural gaps 1, 1, 1, 1, 1, 1,... are constituted by three dependent sets of numbers: (i) CIS of [Incompletely Predictable] odd prime numbers 3, 5, 7, 11, 13, 17,... having CIS of [Incompletely Predictable] prime gaps 2, 2, 4, 2, 4,... plus CFS of solitary [Incompletely Predictable] even prime number 2 having CFS of [Incompletely Predictable] prime gap 1 (ii) CIS of [Incompletely Predictable] even and odd composite numbers 4, 6, 8, 9, 10, 12,... having CIS of [Incompletely Predictable] composite gaps 2, 2, 1, 1, 2, 2,.... and (iii) CFS of solitary odd number '1' [neither prime nor composite]. Treated as Incompletely Predictable problems endowed with "meta-properties", we gave relatively elementary proofs on Polignac's & Twin prime conjectures using these relationships by (1) performing Information-Complexity conservation on prime & composite numbers [and Number '1'] to self-consistently obtain 'Plus Gap 2 Composite Number Continuous Law' for prime gap equal to 2 & 'Plus-Minus Gap 2 Composite Number Alternating Law' for prime gaps greater than 2; and (2) demonstating DA homogeneity with presence of [solitary] cardinality value $\aleph_0$ occurring in all [even number prime gap] subsets of prime numbers and in set of even number prime gaps. **By virtue of the wordings used in these two mentioned Laws; we note that apart from first prime number '2', all prime numbers [represented by prime gap equal to 2 and prime gaps greater than 2] are *dependently* linked to composite numbers [represented by Gap 2 Composite Number].**

Harnassed properties: (1) Nontrivial zeros and two types of Gram points are [*dependently*] derived from "Axes intercept relationship interface" using Riemann zeta function, or its *proxy* Dirichlet eta function; and (2) Prime and composite numbers are [*dependently*] derived from "Numerical relationship interface" using Sieve of Eratosthenes. Using prime gaps as analogy, there are (for instance) "nontrivial zeros gaps" between any two nontrivial zeros with all these gaps of infinite magnitude being Incompletely Predictable entities. Prime number theorem describes asymptotic distribution of prime numbers among positive integers by formalizing intuitive idea that prime numbers become less common as they become larger through precisely quantifying rate at which this occurs using probability. An indirect spin-off arising out of solving Riemann hypothesis result in absolute and full delineation of prime number theorem. This theorem relates to prime counting function which is usually denoted by $\pi(x)$ with $\pi(x)$ = number of prime numbers $\leqslant$ x. In other words, solving Riemann hypothesis is instrumental in proving efficacy of techniques that estimate $\pi(x)$ efficiently. This confirm "best possible" bound for error ("smallest possible" error) of prime number theorem.

In mathematics, logarithmic integral function or integral logarithm li(x) is a special function. Relevant to problems of physics with number theoretic significance, it occurs in prime number theorem as an estimate of $\pi(x)$ whereby its form is defined so that li(2) = 0; viz. li(x) $\equiv \int_2^x \frac{du}{\ln u}$ = li(x) - li(2). There are less accurate ways of estimating $\pi(x)$ such as conjectured by Gauss and Legendre at end of 18th century. This is approximately x/ln x in the sense $\lim_{x \to \infty} \frac{\pi(x)}{x/\ln x} = 1$. Skewes' number is any of several extremely large numbers used by South African mathematician Stanley Skewes as upper bounds for smallest natural number x for which li(x)<$\pi(x)$. These bounds have since been improved by others: there is a crossing near $e^{727.95133}$ but it is not known whether this is the smallest. John Edensor Littlewood, who was Skewes' research supervisor, proved in 1914[8] that there is such a [first] number; and found that sign of difference $\pi(x)$ - li(x) changes infinitely often. This refute all prior numerical evidence that seem to suggest li(x) was always more than $\pi(x)$. The key point is [100% accurate] $\pi(x)$ mathematical tool being "wrapped around" by [less-than-100% accurate] approximate mathematical tool li(x) infinitely often via this 'sign of difference' changes meant that li(x) is the most efficient approximate mathematical tool. Contrast this with "crude" x/ln x approximate mathematical tool where values obtained diverge away from $\pi(x)$ at increasingly greater rate when larger range of prime numbers are studied.

Using classification system in Appendix C, a formula is either non-Hybrid or Hybrid integer sequence. Inequation with two 'necessary' Ratio (R) or equation with one 'unnecessary' R contains non-Hybrid integer sequence. Equation with one 'necessary' R contains Hybrid integer sequence. "In the limit" Hybrid integer sequence approach unique Position X, it becomes non-Hybrid integer sequence for all Positions $\geqslant$ Position X. Kinetic energy (KE) has its endowed units in MJ when m_0 = rest mass in kg and v = velocity in ms^{-1}. In classical mechanics concerning low velocity with v<<c, Newtonian KE = $\frac{1}{2}m_0v^2$. In relativistic mechanics concerning high velocity with v$\geqslant$0.01c, Relativistic KE = $\frac{m_0c^2}{\sqrt{1-(v^2/c^2)}} - m_0c^2$. Obtained from the later by binomial approximation or by taking first two terms of Taylor expansion for reciprocal square root, the former approximates the later well at low speed. We arbitrarily denote inexact DA homogeneity for '<100% accurracy' Newtonian KE and exact DA homogeneity for '100% accurracy' Relativistic KE. "In the limit" Newtonian KE at low speed approach Relativistic KE at high speed, we achieve *perfection*.

Useful analogy: "In the limit" all three versions of Dirichlet Sigma-Power Laws for Gram[y=0] points, Gram[x=0] points and nontrivial zeros as '<100% accuracy' inequations approach *perfection* as '100% accuracy' equations, compliance with inexact DA homogeneity becomes compliance with exact DA homogeneity. We note R1 terms in all inequations contain (2n) and (2n-1) 'base quantities' but these are not endowed with fractional exponent (σ+1) as relevant 'unit of measurement'. Treated as Incompletely Predictable problems, we gave relatively elementary proof of Riemann hypothesis and explain two types of Gram points by using "meta-properties" of relevant Dirichlet Sigma-Power Laws viz. **(1) exact DA homogeneity [occurring when $\sigma = \frac{1}{2}$] in both their equations & inequations and (2) inexact DA homogeneity [occurring when $\sigma \neq \frac{1}{2}$] in both their equations & inequations.**

We define two terms: *perfect symmetry* to denote "even functions" [which are symmetric about vertical y-axis] and "odd functions" [which are symmetric about origin]; and *broken symmetry* to denote "neither even nor odd functions" [which are neither symmetric about vertical y-axis nor origin]. **Then Dirichlet Sigma-Power Laws pertaining to nontrivial zeros (Gram[x=0,y=0] points) in Riemann hypothesis manifest broken symmetry** viz. not satisfying particular symmetry relations present in "even functions" or "odd functions" to **combinedly be classified as "neither even nor odd functions" for both their equations and inequations** whereas **Dirichlet Sigma-Power Laws pertaining to Gram[y=0] points and Gram[x=0] points manifest perfect symmetry** viz. satisfying particular symmetry relations present in "even functions" or "odd functions" to **separately be classified as "even functions" for their inequations and "odd functions" for their equations.**

References

1. Hardy, G. H. (1914). Sur les Zeros de la Fonction $\zeta(s)$ de Riemann. *C. R. Acad. Sci. Paris, 158*, 1012-1014. JFM 45.0716.04 Reprinted in (Borwein et al., 2008)
2. Hardy, G. H.; Littlewood, J. E. (1921). The zeros of Riemann's zeta-function on the critical line. *Math. Z., 10* (3-4), 283-317. http://dx.doi:10.1007/BF01211614
3. Abel, N.H. (1823). Solution de quelques problemes a l'aide d'integrales definies. *Magazin Naturvidensk, 1*, 55-68.
4. Plana, G.A.A. (1820). Sur une nouvelle expression analytique des nombres Bernoulliens, propre a exprimer en termes finis la formule generale pour la sommation des suites. *Mem. Accad. Sci. Torino, 25*, 403-418.
5. Furstenberg, H. (1955). On the infinitude of primes. *Amer. Math. Monthly, 62*, (5) 353. http://dx.doi.org/10.2307/2307043
6. Saidak, F. (2006). A New Proof of Euclid's theorem, *Amer. Math. Monthly, 113*, (10) 937. http://dx.doi.org/10.2307/27642094
7. Zhang, Y. (2014). Bounded gaps between primes, *Ann. Math. 179*(3) 1121-1174. http://dx.doi.org/10.4007/annals.2014.179.3.7
8. Littlewood, J. E. (1914). Sur la distribution des nombres premiers. *Comptes Rendus de l'Acad. Sci. Paris, 158*, 1869-1872.
9. Ting, J (2013). *A228186*. The On-Line Encyclopedia of Integer Sequences. https://oeis.org/A228186
10. Noe, T (2004). *A100967*. The On-line Encyclopedia of Integer Sequences. https://oeis.org/A100967

Appendix A. Gram's Law and traditional 'Gram points'

Named after Danish mathematician Jørgen Pedersen Gram (June 27, 1850 – April 29, 1916), traditional 'Gram points' (Gram[y=0] points) are other conjugate pairs values on critical line defined by $Im\{\zeta(\frac{1}{2} \pm it)\} = 0$. Belonging to Incompletely Predictable entities, they obey Gram's Rule and Rosser's Rule with some characteristic properties outlined by our brief exposition below: Z function is used to study Riemann zeta function on critical line. Defined in terms of Riemann-Siegel theta function & Riemann zeta function by $Z(t) = e^{i\theta(t)}\zeta(\frac{1}{2} + it)$ whereby $\theta(t) = \arg(\Gamma(\frac{(2it+1)}{4})) - \frac{\ln \pi}{2}t$; it is also called Riemann-Siegel Z function, Riemann-Siegel zeta function, Hardy function, Hardy Z function, & Hardy zeta function.

The algorithm to compute Z(t) is called Riemann-Siegel formula. Riemann zeta function on critical line, $\zeta(\frac{1}{2} + it)$, will be real when $\sin(\theta(t)) = 0$. Positive real values of t where this occurs are called 'Gram points' and can also be described as points where $\frac{\theta(t)}{\pi}$ is an integer. Real part of this function on critical line tends to be positive, while imaginary part alternates more regularly between positive & negative values. That means sign of Z(t) must be opposite to that of sine function most of the time, so one would expect nontrivial zeros of Z(t) to alternate with zeros of sine term, i.e. when θ takes on integer multiples of π. This turns out to hold most of the time and is known as Gram's Rule (Law) – a law which is violated infinitely often though. Thus Gram's Law is statement [on the manifested property] that nontrivial zeros of Z(t) alternate with 'Gram points'. 'Gram points' which satisfy Gram's Law are called 'good', while those that do not are called 'bad'. A Gram block is an interval such that its first & last points are good 'Gram points' and all 'Gram points' inside this interval are bad. Counting nontrivial zeros then reduces to counting all 'Gram points' where Gram's Law is satisfied and adding the count of nontrivial zeros inside each Gram block. With this process we need not locate nontrivial zeros but just have to accurately compute Z(t) to show that it changes sign.

Appendix B. Ratio Study and Inequations

A mathematical equation, containing $\geqslant$ one variables, is a statement that values of two ['left-hand side' (LHS) and 'right-hand side' (RHS)] mathematical expressions is related as equality: LHS = RHS; or as inequalities: LHS < RHS, LHS > RHS, LHS $\leqslant$ RHS, or LHS $\geqslant$ RHS. A ratio is one mathematical expression divided by another. The term 'unnecessary' Ratio (R) for any given equation is explained by two examples: (1) LHS = RHS and with rearrangement, 'unnecessary' R is given by $\frac{LHS}{RHS} = 1$ or $\frac{RHS}{LHS} = 1$; and (2) LHS > RHS and with rearrangement, 'unnecessary' R is given by $\frac{LHS}{RHS} > 1$ or $\frac{RHS}{LHS} < 1$. Consider exponent y $\in$ all $\mathbf{R}$ values & base x $\in \mathbf{R} \geqslant 0$ values for mathematical expression x^y. Equations such as $x^1 = x$, $x^0 = 1$ & $0^y = 0$ are all valid. Simultaneously letting both x & y = 0 is an incorrect mathematical action because x^y as function of two-variables is not continuous & is undefined at Origin. If we elect to carry out this "balanced" action [equally] on x & y, we obtain (simple) inequation $0^0 \neq 1$ with associated perpetual obeyance of '=' equality symbol in x^y for all applicable $\mathbf{R}$ values except when both x & y = 0. The Number '1' value in this inequation is justified by two arguments: I. Limit of x^y value as both x & y tend to zero (from right) is 1 [thus fully satisfying criterion "x^y is right continuous at the Origin"]; and II. Expression x^y is product of x with itself y times [and thus x^0, the "empty product", should be 1 (no matter what value is given to x)].

Mathematical operator 'summation' must obey the law: We can break up a summation across a sum or difference but not across a product or quotient viz, factoring a sum of quotients into a corresponding quotient of sums is an incorrect mathematical action. But if we elect to carry out this action equally on LHS & RHS products or quotients in a suitable equation, we obtain two (unique) 'necessary' R denoted by R1 for LHS and R2 for RHS whereby R1 $\neq$ R2 relationship will always hold. We define 'Ratio Study' as intentionally performing this incorrect [but "balanced"] mathematical action on suitable equation [equivalent to one (non-unique) 'unnecessary' R] to obtain its inequation [equivalent to two (unique) 'necessary' R]. Set $\mathbb{C}$ is a field (but not an ordered field). Thus it is not possible to define a relation between two given (z_1 & z_2) $\mathbb{C}$ as z_1 < z_2 since inequality operation here is not compatible with addition and multiplication. But performing Ratio Study to obtain inequations involving $\mathbb{C}$ does not involve defining a relation between two $\mathbb{C}$.

Appendix C. Hybrid method of Integer Sequence classification

Hybrid method of Integer Sequence classification enables meaningful division of all integer sequences into either Hybrid or non-Hybrid integer sequences. Our exotic A228186 integer sequence[9] was published on The On-line Encyclopedia of Integer Sequences website in 2013. It is the first ever [infinite length] Hybrid integer sequence synthesized from Combinatorics Ratio. In 'Position i' notation, let i = 0, 1, 2, 3, 4, 5,..., ∞ be complete set of natural numbers. A228186 "Greatest k > n such that ratio R < 2 is a maximum rational number with R = $\frac{CombinationsWithRepetition}{CombinationsWithoutRepetition}$" is equal to [infinite length] non-Hybrid (usual garden-variety) integer sequence A100967[10] except for finite 21 'exceptional' terms at Positions 0, 11, 13, 19, 21, 28, 30, 37, 39, 45, 50, 51, 52, 55, 57, 62, 66, 70, 73, 77, and 81 with their values given by relevant A100967 terms plus 1. The first 49 terms [from Position 0 to Position 48] of A100967 "Least k such that binomial(2k+1, k-n) $\geqslant$ binomial(2k, k)" are listed below: 3, 9, 18, 29, 44, 61, 81, 104, 130, 159, 191, 225, 263, 303, 347, 393, 442, 494, 549, 606, 667, 730, 797, 866, 938, 1013, 1091, 1172, 1255, 1342, 1431, 1524, 1619, 1717, 1818, 1922, 2029, 2138, 2251, 2366, 2485, 2606, 2730, 2857, 2987, 3119, 3255, 3394, and 3535. For those 21 'exceptional' terms: at Position 0, A228186 (= 4) is given by A100967 (= 3) + 1; at Position 11, A228186 (= 226) is given by A100967 (= 225) + 1; at Position 13, A228186 (= 304) is given by A100967 (= 303) + 1; at Position 19, A228186 (= 607) is given by A100967 (= 606) + 1; etc. Here is a useful concept: Commencing from Position 0 onwards "in the limit" that this Position approaches 82, A228186 Hybrid integer sequence becomes (& is identical to) A100967 non-Hybrid integer sequence for all Positions $\geqslant$ 82.

Appendix D. Tabulated and graphical data on Even-Odd mathematical landscape

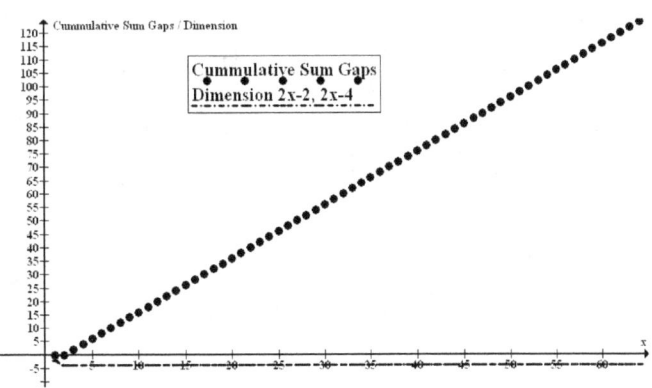

FIGURE 19. Even-Odd mathematical (graphed) landscape using data obtained for x = 1 to 64.

We tabulate in Table 3 and graph in Figure 19 [Completely Predictable] **E-O** mathematical landscape for x = 1 to 64. Involved Dimensions are 2x - 2 & 2x - 4 with Y denoting Dimension 2x - 4 for visual clarity. This mathematical landscape of Dimension 2x - 4 (except for first and only Dimension 2x - 2) will intrinsically incorporate **E** & **O** in an integrated manner. Except for first **O**, all Completely Predictable **E** & **O** and all their associated gaps are represented by countable

TABLE 3. Even-Odd mathematical (tabulated) landscape using data obtained for x = 1 to 64.
Legend: **E** = even, **O** = odd, Dim = Dimension, Y = Dimension 2x - 4 (for visual clarity).

x	E_i or O_i Gaps	ΣEO_x-Gaps	Dim	x	E_i or O_i Gaps	ΣEO_x-Gaps	Dim
1	O1, 2	0	2x-2	33	O17, 2	62	Y
2	E1, 2	0	Y	34	O17, 2	64	Y
3	O2, 2	2	Y	35	O17, 2	66	Y
4	E2, 2	4	Y	36	O17, 2	68	Y
5	O3, 2	6	Y	37	O17, 2	70	Y
6	E3, 2	8	Y	38	O17, 2	72	Y
7	O4, 2	10	Y	39	O17, 2	74	Y
8	E4, 2	12	Y	40	O17, 2	76	Y
9	O5, 2	14	Y	41	O17, 2	78	Y
10	E5, 2	16	Y	42	O17, 2	80	Y
11	O6, 2	18	Y	43	O17, 2	82	Y
12	E6, 2	20	Y	44	O17, 2	84	Y
13	O7, 2	22	Y	45	O17, 2	86	Y
14	E7, 2	24	Y	46	O17, 2	88	Y
15	O8, 2	26	Y	47	O17, 2	90	Y
16	E8, 2	28	Y	48	O17, 2	92	Y
17	O9, 2	30	Y	49	O17, 2	94	Y
18	E9, 2	32	Y	50	O17, 2	96	Y
19	O10, 2	34	Y	51	O17, 2	98	Y
20	E10, 2	36	Y	52	O17, 2	100	Y
21	O11, 2	38	Y	53	O17, 2	102	Y
22	E11, 2	40	Y	54	O17, 2	104	Y
23	O12, 2	42	Y	55	O17, 2	106	Y
24	E12, 2	44	Y	56	O17, 2	108	Y
25	O13, 2	46	Y	57	O17, 2	110	Y
26	E13, 2	48	Y	58	O17, 2	112	Y
27	O14, 2	50	Y	59	O17, 2	114	Y
28	E14, 2	52	Y	60	O17, 2	116	Y
29	O15, 2	54	Y	61	O17, 2	118	Y
30	E15, 2	56	Y	62	O17, 2	120	Y
31	O16, 2	58	Y	63	O17, 2	122	Y
32	E16, 2	60	Y	64	O17, 2	124	Y

finite set of [single] Dimension 2x - 4. Dimensions 2x - 2 & 2x - 4 are symbolically represented by -2 & -4 with 2x - 4 displayed as 'baseline' Dimension whereby Dimension trend (Cumulative Sum Gaps) must reset itself onto this (Grand-Total Gaps) 'baseline' Dimension after initial Dimension 2x - 2 on a permanent basis. Graphical appearances of Dimensions symbolically represented by two negative integers are Completely Predictable with both Even-$\pi(x)$ and Odd-$\pi(x)$ becoming larger at a constant rate. There is a complete absence of Chaos and Fractals phenomena.

Definitive derivation of data in Table 3 is illustrated by two examples for position x = 31 & 32. For i & x ∈ 1, 2, 3, ..., ∞; ΣEO_x-Gap = ΣEO_{x-1}-Gap + Gap value at E_{i-1} or Gap value at O_{i-1} whereby (i) E_i or O_i at position x is determined by whether relevant x value belongs to **E** or **O**, and (ii) both ΣEO_1-Gap and ΣEO_2-Gap = 0. Example, for position x = 31: 31 is **O** (O16). Our desired Gap value at O15 = 2. Thus ΣEO_{31}-Gap (58) = ΣEO_{30}-Gap (56) + Gap value at O15 (2). Example, for position x = 32: 32 is **E** (E16). Our desired Gap value at E15 = 2. Thus ΣEO_{32}-Gap (60) = ΣEO_{31}-Gap (58) + Gap value at E15 (2). Note: in our Dimension (2x - N) system, N = 2x - ΣEO_x-Gap. Then in this unique Dimension (2x - N) system with N = 2x - ΣEO_x-Gap, Dimension (2x - N) when fully expanded is numerically just equal to ΣEO_x-Gap since Dimension (2x - N) = 2x - 2x + ΣEO_x-Gap = ΣEO_x-Gap.

About the Author: Professor John Ting

FIGURE 20. Photo taken in 2016 of author together with his wife and five children.

John Ting is a Researcher on Fundamental Laws of Nature. His novel Hybrid integer sequence A228186 was published in The On-Line Encyclopedia of Integer Sequences in 2013. From 2016 to 2019, he carries out extensive mathematical research with published papers in Number theory on Riemann Hypothesis, Polignac's and Twin prime conjectures. He lives in Australia with his wife and five children. He possesses Medical degree, General Practice qualification, Primary Anesthesia Fellowship Examination and Opioid Replacement license. His work experiences involve the specialty area of Anesthesia, Intensive Care, Pain Medicine, Medicinal Cannabis and Addiction Medicine. His medical publication in 2012 as primary author with the Professor of Nephrology as secondary author include "Supramaximal elevation in B-type natriuretic peptide and its N-terminal fragment levels in anephric patients with heart failure: a case series".

©*Professor Bernhard (Pseudonym) Riemann, viXra, Wednesday 15 January 2020*
Prof. Bernhard (Pseudonym) Riemann, 12 Splendid Drive, Bridgeman Downs, Q4035, Australia
Email: jycting@hotmail.com

www.ingramcontent.com/pod-product-compliance
Lightning Source LLC
Chambersburg PA
CBHW080910220526

45466CB00011BA/3529